Cambridge
checkp•int

**THIRD
EDITION**

Lower Secondary
Science

9

Peter D Riley

**HODDER
EDUCATION**
AN HACHETTE UK COMPANY

To Megan, Harriet and Brea

The Boost knowledge tests and answers have been written by the authors. These may not fully reflect the approach of Cambridge Assessment International Education.

Cambridge International copyright material in this publication is reproduced under licence and remains the intellectual property of Cambridge Assessment International Education.

Every effort has been made to trace all copyright holders, but if any have been inadvertently overlooked, the Publishers will be pleased to make the necessary arrangements at the first opportunity.

Although every effort has been made to ensure that website addresses are correct at time of going to press, Hodder Education cannot be held responsible for the content of any website mentioned in this book. It is sometimes possible to find a relocated web page by typing in the address of the home page for a website in the URL window of your browser.

Third-party websites and resources referred to in this publication have not been endorsed by Cambridge Assessment International Education.

Hachette UK's policy is to use papers that are natural, renewable and recyclable products and made from wood grown in well-managed forests and other controlled sources. The logging and manufacturing processes are expected to conform to the environmental regulations of the country of origin.

Orders: please contact Hachette UK Distribution, Hely Hutchinson Centre, Milton Road, Didcot, Oxfordshire, OX11 7HH. Telephone: +44 (0)1235 827827. Email education@hachette.co.uk Lines are open from 9 a.m. to 5 p.m., Monday to Friday. You can also order through our website: www.hoddereducation.com

ISBN: 978 1 3983 0218 1

© Peter D Riley Ltd 2022

First published in 2005
Second edition published in 2011
This edition published in 2022 by
Hodder Education,
An Hachette UK Company
Carmelite House
50 Victoria Embankment
London EC4Y 0DZ

www.hoddereducation.com

Impression number 10 9 8 7 6 5 4 3 2 1

Year 2026 2025 2024 2023 2022

Cover photo © NicoElNino - stock.adobe.com

Illustrations by Integra Software Services Pvt. Ltd., Pondicherry, India

Typeset in Integra Software Services Pvt. Ltd., Pondicherry, India

Printed in Italy

A catalogue record for this title is available from the British Library.

Contents

How to use this book

To make your study of Cambridge Checkpoint Science as rewarding as possible, look out for the following features when you are using this book:

● These aims show you what you will be covering in the chapter.

Do you remember?

This will show you the ideas you have learnt before. Think about what you already know before you begin.

Science activity

Science activities may be about developing a science skill or making a science enquiry.

Science in context

In this box you will find information about how scientists working alone or together have built up our understanding of the world over time, how science is applied in our lives, the issues it can raise and how its use can affect our global environment.

Science extra
The information in these boxes and any other boxes which have the 'Science extra' heading is extra to your course, but you may find these topics interesting and they may help you with your understanding of the overall chapter topic.

DID YOU KNOW?
This is a fact or piece of information that may make you think more deeply about the topic, or that you may share as a fun fact with your family and friends.

Summary

This box will show you how much you have learnt by the end of the chapter.

This book contains lots of activities to help you learn. Some of the questions will have symbols beside them to help you answer them.
Look out for these symbols:

 This blue dot shows you that you have already learnt some information to help you with this topic.

 If a question has a purple link symbol beside it, you will have to use your skills from another subject.

 This star shows where your thinking and working scientifically enquiry skills are being used.

 Scientists use models in science to help them understand new ideas. This icon shows you where you are using models to help you with your ideas in science.

 This green dot shows you that you are considering a scientific issue in a context which requires your understanding of some of the science facts you have learnt.

 This icon tells you that content is available as audio. All audio is available to download for free from www.hoddereducation.com/cambridgeextras

 There is a link to digital content at the end of each unit if you are using the Boost eBook.

CHALLENGE YOURSELF

These activities are a challenge! You may have to think a bit harder to get the correct answer.

LET'S TALK

When you see this box, talk with a partner or in a small group to decide on your answer.

Work safely

This triangle provides you with extra guidance on working safely.

Words that look like **this** are glossary terms, and you will find definitions for them in the glossary at the back of this book. Other key terms that may not be included in the glossary look like **this**.

Introducing science

In this chapter you will learn:
- how a scientist grows up
- about scientific enquiries
- how models and representations are used in science
- stories about scientists
- about scientists working together (Science extra)
- how people communicate about science.

A scientist grows up

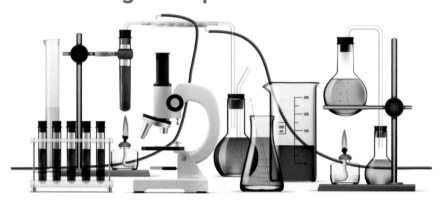

▲ **Figure 1** Scientists use a range of equipment to carry out enquiries.

When do you think people become scientists? Figure 2 on page vii shows a timeline to help you decide.

Babies wouldn't look and listen if they were not curious about their surroundings, so we may say that **curiosity** is one of the first signs of being a scientist.

Later, as children grow and learn to speak, they begin to ask questions about what they see and hear. They want to know how things work – like machines, for example – and why events such as night and day happen.

As children grow, they begin to wonder if they could perhaps make something work better and think creatively about how they could improve it. Later, when they begin making scientific investigations at school, they use their **imagination** and **creativity** to help them to plan their work.

When investigations become more complicated, as they take longer to carry out and more data needs to be collected, students develop **patience** and **perseverance**. For example, when a stop-clock or timer fails to work properly, it can be replaced, and students can begin the experiment again.

Some people believe that everyone behaves like a scientist as soon as they are born. When newborn babies are awake, they watch and listen to the world around them. They make observations.

As babies grow, they pick things up and watch them fall. They perform a simple kind of experiment. Later, they bang objects and listen to the sounds they make. At this time, many activities are investigations.

When children learn more about the world at school, they are struck with a sense of awe and wonder. This makes them to want to find out more.

At high school, enquiries carried out in science laboratories can take the sense of awe and wonder to new heights, as we see how different objects react in controlled situations, or how microscopic organisms swim into view under a microscope, for example.

After learning about biology, chemistry, physics and Earth and space in detail at school, some people develop a great interest in one area of science and want to find out more. They can continue their studies at a college or university and may then go on to work as scientists.

▲ **Figure 2** We are all scientists from a very early age.

LET'S TALK

At what time in their lives do you think people become scientists? Explain your answer.

Which signs of being a scientist do you have? Give examples to support your answer.

---

1 You see a tree fall down in a high wind and hear it crash to the ground. If the tree was a long way away, and nobody could see or hear it, would it still make a sound as it hit the ground? Explain your answer.

2 How could you test your answer to Question 1?

3 Name two signs of being a scientist that you displayed when you answered Question 1.

▲ **Figure 3** Scientific enquiry began 1200 years ago, when Muslim scholars began testing their ideas with investigations.

Scientific enquiry

We make scientific discoveries by carrying out scientific enquiries. A scientific enquiry is divided into three steps and in each step, scientists conduct a range of activities as described here.

Step 1 Purpose and planning

- Use scientific understanding to think up a hypothesis that can be tested.
- Give examples of where unexpected results from enquiries have improved our scientific understanding.
- Use scientific knowledge and understanding to make predictions about what might happen in a scientific enquiry.
- Test the hypotheses by planning a range of different types of investigation to gather appropriate evidence.
- Identify risks in the planned practical work and control them by making a risk assessment.

Step 2 Carrying out a scientific enquiry

- Use observations, tests and secondary sources to sort phenomena, materials, objects and living things into groups. Use classification systems and make **keys** based on the classification system and use them.
- Identify the equipment needed for an experiment or investigation and use it for the purposes for which it was designed.
- Gather more reliable data by increasing the range of observations and measurements made and, where necessary, increase the number of times they are repeated.
- Explain why accurate and precise measurements are needed in an experiment or investigation in order to produce reliable data.
- Always consider the risks involved in practical work and use a risk assessment to work safely.
- Use scientific knowledge and understanding to decide whether to gather evidence by first-hand experience, through experiments and observations, or whether to use secondary sources to provide the data that is needed.
- Collect the number of observations and measurements you need to provide reliable data and record them. Summarise them appropriately for further examination.

▲ **Figure 4** Scientists choose and assemble equipment once they have decided what evidence needs to be collected and what observations need to be made.

Step 3 Analysis, evaluation and conclusions

- Look at the amount of evidence that has been collected and assess its strengths to help you decide whether it proves or disproves the hypothesis
- Look for patterns and trends in results and, if found, describe them. Identify any anomalous results and suggest why they do not fit in with a pattern or trend.

- Look through the results and decide what they might show. Then consider the limitations of the data before making a conclusion. Follow the conclusion with suggestions for further experiments and investigations to make the data more reliable and less limited.
- Present the results in such a way that they can be easily interpreted. If they form a set of data points, predict results between them.
  A data point is a single piece of information. In science, it could be a measurement of temperature, such as 37 °C, or a measurement of **force**, such as 10 N. When you record the temperature over a period of time, you make a collection of data points in a table, although they just look like numbers. When you present your data as a graph, the temperature at a particular time forms a point – a data point on your graph.
- Look at the experiments and investigations that have been made by yourself and others and evaluate them in terms of providing reliable data for conclusions. From the evaluation, suggest how the work can be improved and explain any changes that you think need to be made.

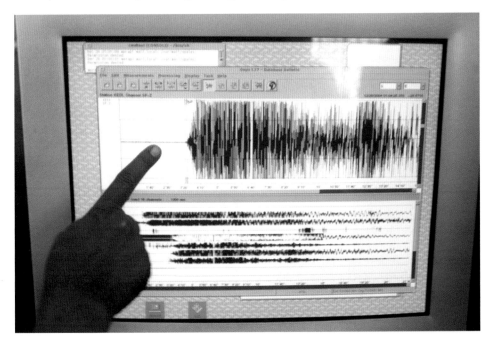

▲ **Figure 5** A seismograph is used to measure the **vibrations** of the Earth's surface. This one is showing vibrations due to an earthquake which measured over 9 on the Richter scale. This scale is named after Charles Richter, who invented it to measure the energy released by earthquakes.

A closer look at scientific enquiry

You should be familiar with these scientific terms:

- hypothesis
- prediction
- investigation
- experiment
- variables
- fair test
- trend
- pattern
- anomalous result
- data/data points.

Using models and representations in science

When data collected in a scientific enquiry is analysed, evaluated and conclusions are drawn, scientists sometimes make a model or use an **analogy** so that the data and conclusions are easier to understand. Models have strengths and limitations. You should be able to recognise and describe them. However, models and analogies are built on the results of current investigations, and conclusions can change as further investigations are carried out, and the models and analogies can be changed too. Some models, however, do remain the same, such as the symbols in chemistry and the formulae used throughout science studies. You should learn to use these as you work though this book and move onto further studies.

Unexpected results

When an enquiry begins, a hypothesis is made, followed by a prediction based upon it. The investigation takes place and usually the scientists have some idea of what to expect based on their knowledge and understanding. Sometimes the investigation leads to a result that is unexpected. When this happens, the scientists gain some more knowledge and understanding of a process, phenomenon or event and this can lead to further investigations. Try the following enquiry and see whether or not your result is the one you expected.

What happens when the ice melts?

You will need:

a glass, a piece of ice or ice cubes, a jug of water, a camera.

The question can be more precisely set out as follows:

When a glass of water is full to the brim and has a piece of ice in it, what happens to the water in the glass as the ice melts?

Use this version of the question to construct a hypothesis.

Hypothesis

Construct a hypothesis in which you say what will happen and explain why it will happen.

Prediction

State your prediction based on your hypothesis.

Process

1. Fill a glass almost full of water.
2. Add a piece of ice to the water.
3. Place the glass somewhere that it can be safely left for the ice to melt.
4. Fill the glass to the brim with water.

▲ **Figure 6** What happens when the ice melts?

5 Photograph the glass and the table or benchtop on which it stands.

6 Leave the ice to melt.

7 When the ice has completely melted, take a second photograph.

Examining the results

Compare the data in the two photographs.

Conclusion

Compare your evaluation with your hypothesis and prediction and draw a conclusion.

An explanation of the result can be found on page 87, but to fully understand it you should read Chapters 7 and 8. There are also several more examples of unexpected results leading on to discoveries in other chapters of this book.

Stories about scientists

Here are three stories about people from the past showing signs of being a scientist as they grew up. They later went on to make important discoveries. The first person lived about a thousand years ago, while the second and third lived and made important investigations in the twentieth century.

Alhazen's first investigations on light

Alhazen (also known as Abu Ali al-Hasan ibn al-Hasan ibn al-Haytham) was a Persian scientist who lived about a thousand years ago. He read about the ideas of the Ancient Greeks and became interested in many subjects such as astronomy, physics, mathematics, medicine and engineering.

He discovered that the Ancient Greeks believed that we see objects around us because light rays leave the eye and hit the objects, making them visible. Alhazen thought about all the objects that we can see and then considered the stars. When he looked up at the sky, shut his eyes and then opened them, he could see the stars immediately. He knew that the stars were a long way away and believed they were just too distant for rays from the eyes to reach them in an instant.

Alhazen believed that the reverse of the Ancient Greeks' idea was true – that light comes from the objects we see. He also believed that rays of light travel in straight lines. He set about testing his idea by setting up a dark room with a small hole in one wall. In a dark area outside the room he hung two lanterns, and then went back inside. He saw two spots of light on the opposite wall to the hole. He reasoned that each spot came from one of the lanterns, and when he checked their positions, he found that each lantern, hole and spot were in a straight line.

▲ **Figure 7** Alhazen was particularly interested in investigating how light travels.

He then covered each lantern in turn with a cloth and discovered that the spot of light it produced became darker, but when he removed the cover the spot became bright again. From this, he reasoned that light does not come from the eye, but from objects around it that produce light.

The evidence provided by Alhazen's work was later used by scientists in Europe to build up our knowledge about light.

4   From looking at Alhazen's interests, would you say that he was curious? Explain your answer.

5   In Alhazen's investigation, do you think the cloth with which he covered the lanterns was opaque or translucent? Explain your answer.

6   What evidence did Alhazen first consider when thinking about light?

7   What creative thought did Alhazen apply to the evidence?

8   What do you think Alhazen decided he needed in order to test his ideas?

9   Why do you think Alhazen used two lanterns?

Anna Mani and the study of the weather

Earlier we saw how curiosity, imagination, creativity and perseverance are needed for success as a scientist. In this story, we see how they played an important part in the life of Anna Mani.

Anna Mani (1918–2001) was born in India. She loved reading, and on her eighth birthday, in 1926, she asked for a set of encyclopaedias as a gift. By the time she was 12, she had read all the books in her local library. Mani thought about becoming a doctor, but found that she particularly enjoyed physics, so she went to college in Madras to study it. In one set of investigations, she studied how diamonds and rubies absorb light and what happens to this light. This involved exposing photographic film to the gemstones for up to 20 hours. During this time, Mani did not leave her experiments but slept in the laboratory with them!

Mani graduated in 1939 with a BSc Honours degree in chemistry and physics. In 1940 she won a scholarship to the Indian Institute of Science in Bangalore. In 1945 she travelled to England and continued her study of physics at Imperial College, London. This time she studied how to improve the accuracy of **meteorological** instruments such as thermometers, rain gauges and anemometers. In 1948, Mani returned to India and brought together scientists and engineers to make both instruments that she had studied before as well as instruments that record their own data.

From her studies on the weather, Mani became interested in the idea of using the energy in sunlight and wind as a source of power to generate electricity. In 1957–8 she set up stations around India to measure how solar **radiation** varied with the seasons. This data was then used to plan ways to capture energy from sunlight. Mani also organised the collection of data on the wind from over 700 weather stations across India, which has since been used to plan the setting up of wind farms. In around 1960, her weather studies also led her to become interested in the ozone layer, and she devised equipment to measure the amount of ozone in the atmosphere.

▲ **Figure 8** Anna Mani was best known for her work as a meteorologist.

In addition to her studies, Anna Mani enjoyed the company of other scientists and the development of science in her country and around the world. She was a member of the Indian National Science Academy, the International Solar Energy Society and the American Meteorological Society. In 1976, she retired from her post as deputy director general of the Indian Meteorological Society, and in 1987 she was awarded the INSA KR Ramanathan medal for her achievements in science.

Mani, like most scientists, had other interests outside her work which helped her to relax, and she particularly liked trekking and bird-watching.

10 Did Anna Mani show signs of being curious when she was young? Explain your answer.

11 Patience and perseverance show dedication to an activity. When did Mani show great dedication to her work?

12 At what stage of her life was Mani's work mainly involved with obtaining and presenting evidence?

13 At which point in her career did Mani's work involve creativity on a large scale?

14 What sources of secondary data did Mani's work provide?

15 Use the information in the text to draw a timeline of the key events in Anna Mani's life.

Trevor Baylis and the clockwork radio

Trevor Baylis (1937–2018) was born in London, England, and became a swimmer, taking part in international competitions by the age of 15. He later got a job in a laboratory and also studied engineering at a local college. He went on to apply his scientific skills in the development of swimming pools. He combined his love of swimming and his work in swimming pool development by providing entertainment through swimming and working as a stunt man in a glass-sided pool.

In 1989, Baylis watched a television programme about **AIDS** in Africa. The programme stated that it was believed that the spread of the disease could be slowed if people could be educated about it by means of radio programmes. The problem was that radios needed either batteries or a power supply – and millions of people at risk of getting AIDS were too poor to afford batteries, or lived in villages in the countryside that did not have a power supply. Baylis immediately thought about a way to solve the problem by inventing a radio that did not need an outside source of electricity.

▲ **Figure 9** Trevor Baylis invented a radio that needs no batteries or mains electricity.

CHALLENGE YOURSELF

Create a presentation to explain how the work of Trevor Baylis helped so many people.

CHALLENGE YOURSELF

If you were to be a scientist, what would your science story be? Look at your interests as a child growing up, your interests in science and school, and any topics in biology, chemistry, physics or Earth and space.

First, he connected a radio to an electric motor, and then connected a handle to the electric motor. When an electric motor receives a supply of electricity, it turns, but if the motor is turned with a handle, it generates electricity. Baylis checked that when he turned the handle the motor made electricity and the electrical energy made the radio work.

The problem was that he had to keep turning the motor to make the radio work. He then thought of a way to store the energy from turning the handle so he could listen to the radio without having to make it work at the same time. He used a clockwork spring. It was wound up by turning the handle and then allowed to unwind, and as it did so, its stored energy passed to an electrical generator, which then supplied electricity to make the radio work. The first version of his clockwork radio could play for 14 minutes after the spring had been wound up for two minutes. Since then, wind-up radios have been developed that can charge up rechargeable batteries and some also have solar cells.

16 How did Trevor Baylis show that he was curious about different things in his early life?

17 Name two things in Baylis' life, not connected with science, that eventually led to him caring for the disabled.

18 What stimulated Baylis into devising the clockwork radio?

19 What question might Baylis have asked himself when thinking about his new radio?

20 What question do you think Baylis was finding the answer to when he used the handle to turn the motor connected to the radio?

21 What question do you think Baylis was finding the answer to when he used the clockwork spring?

LET'S TALK

Do you think it is a good idea for scientists to have other interests besides science, as Anna Mani did? Explain your answer.

Do you think it is a good idea for scientists to have an interest in science subjects other than the one they are working in, like Alhazen? Explain your answer.

Science extra: Scientists working together

In the past, scientists such as Alhazen, Galileo, Newton and Boyle worked alone or with a few assistants. In more recent years, very few scientists work alone. Today, most scientists work with others in their own country or other countries across the world, just as Anna Mani did, and sometimes they form teams to study a particular subject.

▲ **Figure 10** Scientists working together.

However, scientists have been coming together to work and share their enthusiasm for science for hundreds of years. In 1603 in Rome, four young scientists set up the Accademia dei Lincei, which means 'academy of the lynx-eyed'. At that time, people believed that the lynx was so keen-eyed, it could see through rocks and trees, so the name suggests that the scientists could use their powers of observation to see things that others could not.

Later, in 1660, the Royal Society of London for Improving Natural Knowledge was formed, and today it is called the Royal Society. Its Latin motto is *Nullius in verba* which can be translated as 'take nobody's word for it'.

▲ **Figure 11** The Eurasian lynx.

LET'S TALK

If you were to form a science club at school, what would you use as your inspiration – an animal like the lynx, a famous scientist, the activity of doing an experiment or something else? If you were to make a logo for your club, what would it be? What motto could your club have? It could be in any language, not just Latin or English.

Communicating science

As you have studied over the years, you have been encouraged to communicate your ideas. They way that you did this may have started with drawing pictures of your observations or experiments and then may have moved on to recording data in tables, charts and graphs, or to making and demonstrating models.

All these are features of communication, which is essential in the work of scientists, as they share and peer review their work in scientific societies like the Royal Society and others around the world. These features of communication are used by scientists to inform everyone about the latest scientific developments and what they mean for our lives. The first organisation to do this was set up in 1799 in London, and was called the Royal Institution. It was formed to communicate science and technology to anyone who wanted to come and attend their lectures. The Royal Institution thrives to this day, giving lectures every year in late December, and taking versions of their lectures around the world. You can watch their lectures and other short films online on their YouTube channel.

LET'S TALK

Do you remember making a presentation about science? What was the topic and what did you do? How did you feel about making the presentation and what feedback did you receive? Could you see yourself as a science communicator? Explain your answer.

▲ **Figure 12** Communicating ideas is essential in the work of all scientists, even young ones who are just beginning their formal scientific education.

Summary

✔ Investigating our surroundings occurs from an early age and can lead to developing an interest in science.
✔ There are three stages in scientific enquiry.
✔ We use scientific models to make data and conclusions easier to understand.
✔ Studying the lives of scientists reveals the human qualities that are needed to make scientific discoveries, and shows how scientific knowledge is developed through collective understanding and scrutiny over time.
✔ Communication is essential to help scientists work together, and to inform all people of the developments in science and their consequences.

End of chapter questions

1 Plants grow towards the light. There are different colours in white light.

Work out a testable hypothesis to further extend your knowledge about plant growth and light from these two pieces of information.

2 As the Sun moves, it casts shadows of different lengths and directions. The Sun also takes many hours to cross the sky.

Work out a testable hypothesis to further extend your knowledge about the movement of the Sun and the measurement of time.

3 What do you predict will happen when a magnet is put on one side of a piece of cardboard and a steel paper clip is put on the other? Explain your prediction.

4 What do you predict will happen to a puddle on a sunny day? Explain your prediction.

5 A plant grows the same amount each day. Its height is measured and found to be 10 cm. Five days later it is found to be 20 cm high.

a Use these data points to produce a line graph.

b If you had measured the plant height after three days, what would it have been?

6 The number of birds coming to roost in a tree over five consecutive days is recorded in Table 1.

▼ **Table 1**

Day	Number of birds
Sunday	3
Tuesday	35
Thursday	7
Saturday	16
Monday	27

a Construct a graph with the x-axis showing the days and the y-axis showing the number of birds.

b Plot the data points on the graph to make a scattergram.

c Look at the arrangement of the data points and decide where you could draw a line through them from Sunday to Monday that suggests a trend. This is called the line of best fit.

d Using this line of best fit, predict the number of birds coming to roost on Wednesday.

e How reliable is your prediction? Explain your answer.

1 Water and life

In this chapter you will learn:
- how water is absorbed from the soil by root hairs
- how water is transported through plants in xylem vessels
- about different types of stems (Science extra)
- how water that evaporates from the leaves is replaced by water from the xylem vessels in a process called transpiration
- how water absorbed from soil contains mineral salts that are used by plants
- how plants use magnesium and nitrates
- about monitoring vegetation with satellites (Science in context)
- how plant cells store water and minerals (Science extra)
- to describe the human renal system
- how kidneys remove urea from the blood
- about kidney machines (Science in context).

Do you remember?

- Water exists in three states of matter. What are they?
- Is water an element, a compound or a mixture?
- Excretion is a **characteristic** of life. Name five more.
- What does the term 'excretion' mean?
- Name seven parts of a typical plant cell, such as one found in a leaf.
- Describe a root hair cell.

The watery planet

Voyager 1 was a **space probe** that was launched from Earth in 1977 with the mission to leave the solar system and travel through interstellar space. In 1990, when it was six billion kilometres from our planet, it sent back a photograph which shows the Earth as a blue spot about the size of a full stop.

The Earth appears blue because of the water on its surface. Sunlight can be divided into different colours, as Newton showed almost four hundred years ago. These colours are absorbed by the water as the light rays pass down into it. The last of the colours to be absorbed is blue, and this is what made the planet look blue in Voyager's picture. For spacecraft that do not travel as far away, blue is still the main colour in photographs that are taken of Earth, but there are other colours too, which come from the land surface.

The main colour reflected into space from the land is green, which comes from the plants that grow there. Plants obtain water through the water cycle and without it they, like all life on the planet, would not exist.

> **LET'S TALK**
> Use the water cycle to explain how water from the surface of the oceans becomes available for the plants on the land.

▲ **Figure 1.1** The Earth as photographed from about 1.6 million kilometres away.

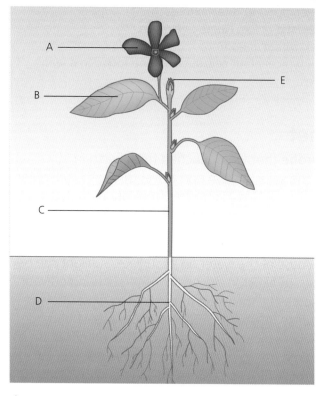

▲ **Figure 1.2** The organs of a flowering plant.

Organs of a flowering plant

There are five main organs in the body of a flowering plant. They are the root, stem, leaf, flower and bud.

The transport of water through the root and root hairs

Most plant roots have projections called root hairs. The tips of the root hairs grow out into the spaces between the soil particles. There may be up to 500 root hairs in a square centimetre of root surface. They greatly increase the surface area of the root so that large quantities of water can pass through them into the plant. The water in the soil is drawn into the plant to replace the water that is lost through **transpiration** from the surface of the leaves. The plant does not have to use energy to take the water in.

1 Match the names of the organs to the letters in Figure 1.2.

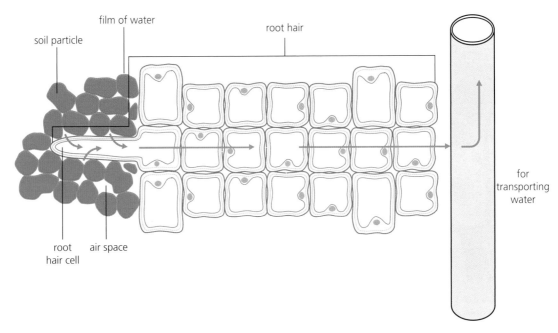

▲ **Figure 1.3** Schematic drawing of the movement of water in the root of a plant.

Xylem

Plants have cells which form tubes to transport water. The cells form columns in the plant and when they die, the walls between them break down to form tubes as shown in Figure 1.4.

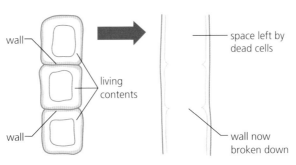

▲ **Figure 1.4** How a tube for transporting water is made in a plant.

▲ **Figure 1.5** Celery stalks on a celery plant.

Each water-conducting tube is called a **xylem** (pronounced *zylem*) vessel. A group of vessels form xylem tissue, which makes up a part of structures called vascular bundles. These run through the plant from the root to the leaf, where they form leaf veins. If you have ever eaten celery, you will be familiar with vascular bundles. They form the celery fibres which sometimes stick between your teeth! The celery stalk is sometimes used in experiments.

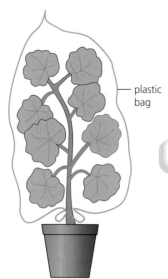

plastic bag

▲ **Figure 1.6** Equipment used to show transpiration.

CHALLENGE YOURSELF

Set up a plant like the one shown in Figure 1.6. Use one of the plants from the first challenge in this chapter if possible. Leave the plant in sunlight for a day then examine the bag for water. What do you find? What chemical could you use to test for water?

CHALLENGE YOURSELF

Select two data points from your result and construct a line graph. Use it to predict the result half way between the data points.

Transport of water through the stem and leaves

The stem

The stem supports the leaves, and if a leafy plant is enclosed in a clear plastic bag like the one shown in Figure 1.6, after a while water is found inside the bag.

Can you track the path of water through a stem?

You will need:

two similar-sized celery stalks (one with plenty of leaves and one with all the leaves removed), a beaker, water mixed with blue ink, a scalpel, a dissection board.

Hypothesis

Construct a testable hypothesis to explain what might happen if a leafy stem and a stem without leaves were dipped into coloured water and left for a while. Explain your hypothesis.

Prediction

You are to use two celery 'stems' – one with leaves and one without. Make a prediction about the outcome of the experiment.

The path of water through a stem can be traced by cutting open the stem and looking for where the blue ink in the water has reached.

Planning and investigating

Make a plan using this information, stating what you will record and for how long. If your teacher approves, try it.

Examining the results

Compare the height of the ink and water mixture in both celery stalks.

Conclusion

Compare your analysis with your prediction. Does the evidence support or refute the prediction?

Explain how your conclusion could have limitations.

How could further investigations be made to see if you could establish a trend or pattern in the data?

2 In Figure 1.6, where did the water in the bag originally come from? The leaves, the stem or the soil? Explain your answer.

Science extra: Types of stem

The stems of most plants grow upwards and support themselves. Some plants have stems that are too weak to support themselves, and instead grow up the sides of other, larger plants. They may have structures called tendrils, which look like little springs, growing and curling around the parts of the supporting plant.

A few plants have stems that grow across the ground or even under it. The ginger plant has an underground stem in which it stores food.

▲ **Figure 1.7** The **tendril** of this pea plant is wrapping around the stem of another plant.

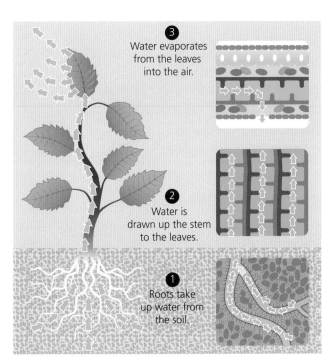

❸ Water evaporates from the leaves into the air.

❷ Water is drawn up the stem to the leaves.

❶ Roots take up water from the soil.

▲ **Figure 1.8** The transpiration stream.

The leaf

When water evaporates from cells in the lower layer of the leaf, water vapour forms. If there is less water vapour outside the leaf than inside it, the water vapour will diffuse out through the holes on the lower side. This makes the cells in the lower layer short of water, so water will move from the xylem tissue in the veins into the cells. The water lost in the veins is replaced by water passing up the xylem tissue in the stem and root.

The process by which plants lose water from their leaves is called **transpiration** and the movement of water from the roots through the stem to the leaves is called the transpiration stream.

Can water be shown to escape from leaves?

You will need:

a leafy shoot in a pot or part of a branch attached to a tree, a clear plastic bag, sticky tape or string to tie up the opening of the bag, anhydrous copper sulfate.

Hypothesis

If something is releasing water vapour in an enclosed space, the vapour may turn to liquid on the walls. Use this piece of information and a clear plastic bag to work out a testable hypothesis to answer the enquiry question.

Prediction

Write a prediction based on your hypothesis.

Planning, investigating and recording data

Work out a plan to test your hypothesis and, if your teacher approves, try it and record your results.

Examining the results

Examine the results and compare it with your prediction.

Conclusion

Decide whether your investigation answers the question and draw a conclusion.

What are the strengths and limitations of your investigation?

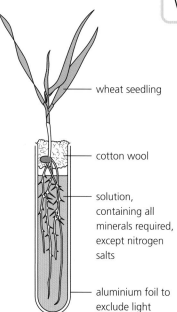

wheat seedling

cotton wool

solution, containing all minerals required, except nitrogen salts

aluminium foil to exclude light

▲ **Figure 1.9** This plant has been set up in a solution from which nitrogen salts have been omitted.

Minerals and how plants use them

The water that enters the plant is not pure. It is a solution of mineral **salts**. The mineral salts are carried in the water as it passes up the plant. They take the same path as the water. Mineral salts can come from rocky fragments in soils, or from organic matter or from natural cycles such as the nitrogen cycle.

Two examples of minerals in the soil are nitrogen – in the form of nitrate salts – for making proteins, and magnesium to make chlorophyll.

Scientists found out about the importance of each mineral in the following way. They made a solution of salts with all the minerals that plants need, except the one they were investigating. For example, if the effect of nitrogen was being investigated, they made a solution that contained all the minerals except nitrogen. The plant was then grown for a few weeks in this solution (Figure 1.9), alongside a plant growing in a solution containing all the mineral salts required, and the differences between the two plants were recorded. From these studies, scientists discovered the following:

- Nitrogen is needed for the development of the leaves. Without nitrogen, the leaves turn yellow and the plant shows poor growth. Further studies have shown that nitrogen is needed for making the green pigment chlorophyll and for making proteins that form part of the structure of the plant.
- Magnesium is needed to make chlorophyll. Without magnesium, the chlorophyll in the existing leaves begins to break down and areas around the veins become yellow.

▲ **Figure 1.10** Plants showing mineral deficiency of nitrogen and magnesium.

3 Why might a plant show poor growth?

4 When investigating the importance of different mineral salts, why was a solution used for each experiment, rather than a mixture of soil and the solution?

Science in context

Monitoring vegetation with satellites

The conditions of the soil, such as its water and nitrogen content, can be observed and measured using **satellites** which send back images of the land being examined. The conditions of the growing crops can also be observed and measured, so that farmers can monitor the conditions on their farms and plan their activities to collect the largest yield from their crops.

▲ **Figure 1.11** The images are coloured artificially to show the conditions in the field of crops. By studying the colours, scientists can advise on how to treat the crops to produce their greatest yield.

Images produced by satellites are also used to monitor conditions in **ecosystems**, including in forests and in the oceans. These images help scientists to plan **conservation** strategies and to collect data to use in studies on climate change.

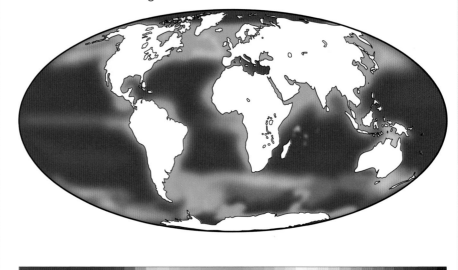

| 0.03 | 0.1 | 0.3 | 1 | 3 | 10 | 30 |

▲ **Figure 1.12** The primary **producer** in the ocean ecosystem is phytoplankton. This map shows their worldwide distribution. The green areas show where they are mostly found, the pink areas show where the fewest phytoplankton are found, and blue areas shows where they occur in small numbers.

5 Describe the distribution of phytoplankton in the world's seas and oceans.

6 Phytoplankton grows best where there are high concentrations of nitrogen and phosphorus in the water. Where are these waters found?

> ### Science extra: Water and minerals in cells
>
> In the centre of plant cells is a space called the **vacuole** through which water and minerals pass. Not all the water and minerals leave these cells, as they must have a certain amount of water inside them to keep themselves alive. In animal cells, which do not have the large vacuole of plant cells, the water and minerals are present in the cytoplasm.
>
>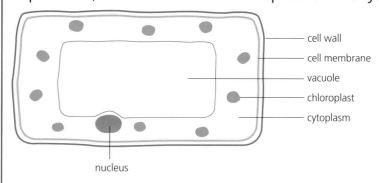
>
> ▲ **Figure 1.13** A typical plant cell.
>
> Most scientists believe that the water and concentration of its minerals in the cells of living things today are similar to the seawater of ancient times, in which life is believed to have evolved.

The human renal system

The water in all our body cells is regulated by the activity of the **renal system**. This system also removes a harmful substance called **urea** by releasing it from the body as a watery solution called urine. Sometimes the renal system is also called the **urinary system**.

The renal system is part of the body's **excretory system**, which also includes the lungs excreting carbon dioxide and the skin excreting a tiny amount of urea in sweat.

Cells need a certain amount of a range of substances, such as oxygen and glucose, **amino acids** from digested proteins to make many of the cells' structures, and vitamins and minerals to keep the body healthy. Cells receive these substances from the blood as it passes by them in capillaries. As energy is released in **respiration** in the cells, carbon dioxide is also produced which is taken up by the blood plasma and excreted in the lungs. Although the body needs amino acids for the growth and repair of its cells, it cannot store them. This means that any amino acids that the body does not use must be destroyed and this process takes place in the liver. In this process, urea is produced and passes into the plasma. Urea is a poison and is excreted from the body in the kidneys.

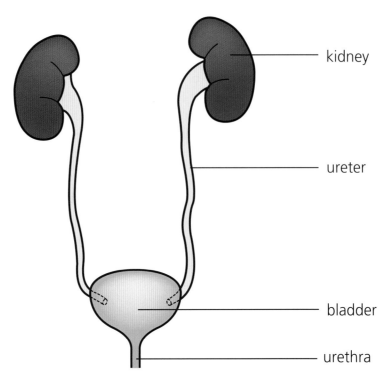

kidney

ureter

bladder

urethra

▲ **Figure 1.14** The kidneys and the other structures involved in the excretion of urine.

Inside the kidney

A kidney is a blood filter. You will have used the filtering process earlier in your science studies and will use it again in this book, on pages 103 and 104. If you look at the diagrams on those pages (Figures 10.7 and 10.9 respectively), it may help you to refresh your memory about filtering substances.

Inside a kidney, the liquid that is filtered is blood. In grade 8, you learnt that the blood has three components: the red blood cells, the white blood cells and the plasma. The plasma contains lots of substances dissolved in it, including urea. When the blood enters the kidney, it passes through lots of small tubes, and some of them are arranged to act as filters and remove the urea in a watery liquid called urine. The cells and the other substances remain in the plasma, and continue to pass around the body.

DID YOU KNOW?
In a healthy adult human kidney, there can be up to 1.5 million nephrons.

The kidney machine

The working of the kidney is quite complex and requires cells that use energy to bring vital substances back into the blood after filtration has occured to remove urea. If the kidneys do not work properly due to disease, a person must use a kidney machine which removes the urea, yet keeps the other vital substances in the blood. Kidney machines

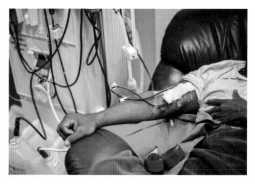

▲ **Figure 1.15** A patient receiving dialysis.

work using a process called **dialysis**, which is a combination of diffusion and filtration to remove the harmful substances from the blood. It is an example of how scientific knowledge and understanding can be used to engineer a piece of technology which has saved many lives. Most people who use a kidney machine need to use it three times a week. Each filtration session takes about four hours.

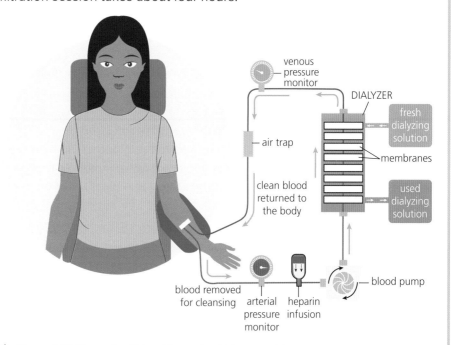

▲ **Figure 1.16** The path of blood through a kidney machine.

Summary

✔ Water is absorbed by root hairs in plants.
✔ Water is transported through plants from the roots to the leaves in tubes called xylem vessels.
✔ The movement of water through a plant and its **evaporation** from the leaves is called transpiration.
✔ Mineral salts are absorbed and transported to the different parts of a plant in water.
✔ Plants require magnesium to make chlorophyll and nitrogen to make protein.
✔ The kidneys, which are part of the human renal system, filter the blood to remove urea.
✔ Urea is removed from the human body in urine.
✔ The use of satellites to monitor water and mineral salts in the soil can help people who are trying to protect the **environment**.
✔ Scientific knowledge about the function of human kidneys was used to develop the kidney machine, which helps people whose kidneys don't work properly.

End of chapter questions

1 What are the projections on a root called?
2 Which part of the plant connects the root to the leaves?
3 Name two minerals that plants need to stay healthy. What does each mineral do?
4 What is the transpiration stream?
5 What happens to the blood in the kidney?
 6 Imagine you are a water molecule that has entered a root. Describe your journey through a plant until you leave through a leaf.

 Now you have completed Chapter 1, you may like to try the Chapter 1 online knowledge test if you are using the Boost eBook.

Photosynthesis

In this chapter you will learn:
- to explain the process of photosynthesis and know where in the plant it takes place
- about the willow tree experiment (Science in context)
- about plants and the air (Science in context)
- what plants need in order to produce starch
- about the discovery of oxygen (Science in context)
- the word equation for photosynthesis
- about the perfect environment for plant growth
- how we are using scientific knowledge to grow more food (Science in context).

Do you remember?

- Where do plants get their energy from?
- What do we call the structures in a plant cell that release energy in a controlled way?
- Identify the features of the plant cell shown in Figure 2.1.
- What is the word equation for aerobic respiration that takes place in plant cells?

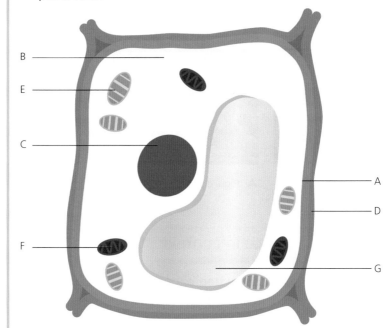

▲ **Figure 2.1** Structures in a plant cell.

Photosynthesis

Photosynthesis is the process by which plants make **carbohydrates** for food. In this chapter, you will build up more detail about photosynthesis by following a research programme.

Scientists set up research programmes to try and make their discoveries in an orderly way. They do not just make one enquiry; they look at the work of other scientists as well as their own earlier work and use this for setting up a series of enquiries. Sometimes, these may serve to support previously drawn conclusions or show that they were wrong. This chapter is set out as a research programme, explaining what is meant by photosynthesis. It also gives you an idea of how a simple model produced by one enquiry can be modified by others to produce a more detailed model of photosynthesis.

Plants grow over most of the land on the planet. The regions of greatest plant growth are the rainforests, but even in deserts or in the cold regions around the North Pole, plants manage to put down their roots and grow.

▲ **Figure 2.2** A rainforest.

▲ **Figure 2.3** The African desert.

▲ **Figure 2.4** Willows growing in the frozen tundra.

Science in context

The willow tree experiment

People have probably known from the earliest times that plants need water to survive and grow, but the scientific investigation to find out how plants grow began with an experiment on a willow tree.

In the seventeenth century, Belgian scientist Joannes Baptista van Helmont (1580–1644) performed an experiment on a willow tree.

He was interested in what made it grow. At that time, scientists believed that everything was made from four 'elements': air, water, fire and earth. Van Helmont believed that water was the most basic 'element' in the universe and that everything was made from it. He set up his experiment by weighing a young willow plant (a sapling) and the soil it was to grow in. Then he planted the sapling in the soil and provided it with nothing but water for the next five years. At the end of his experiment, he found that the tree had increased in **mass** by 73 kilograms, but the soil had decreased in mass by only about 60 grams. He concluded that the increase in mass was due to the water the plant had received. If we were to summarise his conclusion, it could look like this:

water → mass of plant

▲ **Figure 2.5** Watering a willow tree.

1   How fair was the willow tree experiment? Explain your answer.
2   Did the result of the experiment support van Helmont's beliefs? Explain your answer.
3   van Helmont provided data about the mass of the tree and the mass of the soil. What other data could he have recorded?
4   If you were to repeat the experiment, how would you improve it?

There was a great increase in making new scientific discoveries during the sixteenth century. This time was called the **Scientific Revolution**, and van Helmont was born during this time and was inspired by the scientists who were developing ideas and experiments. As the scientists worked, some results were unexpected and led to improved scientific understanding. Different types of investigations were tried to test the hypotheses that the scientists constructed. These ideas changed over time and led to other, new experiments which were inspired by older work. The work of Stephen Hales is an example which built on the work of van Helmont.

Science in context

Plants and the air

Stephen Hales (1677–1761), an English scientist, discovered that 'a portion of air' helped a plant to survive, and Jan Ingenhousz (1730–99), a Dutch scientist, showed that green plants take up carbon dioxide from the air when they are put in the light. By this time, it was also known that water contains only the elements hydrogen and oxygen, while carbohydrates contain carbon, hydrogen and oxygen. All this information led to a review of van Helmont's idea that only water was needed to produce carbohydrates. The review began by considering what else was around the plant apart from water. It was known from van Helmont's work that the soil contributed only a very small amount to the increased mass of the plant. The only other material coming into contact with the plant was the air. Ingenhousz's work suggested that the carbon dioxide in the air was important. You can test this idea in the laboratory today.

▲ **Figure 2.6** Stephen Hales.

▲ **Figure 2.7** Jan Ingenhousz.

5 Did Hales perform his experiments at the same time as van Helmont? Explain your answer.

6 How had the idea of elements changed from the days of van Helmont to the time of Hales and Ingenhousz?

7 Why did a review of van Helmont's work lead to the idea that air could be important in making food?

8 How did Ingenhousz's discovery suggest that carbon dioxide might be important in making food?

9 Which two activities in step 3 of making a scientific enquiry (see pages viii–ix) were being used in reviewing van Helmont's work?

After studying a little of the history of science, it is now your turn to make discoveries. But first you need to review the information you have gathered.

Earlier in the chapter, you saw that the work of van Helmont showed that water is needed for the plant to grow. We know that if plants do not receive water, they die. We also know that if plants are kept in the dark, they will try to find a way to the light. Without light, plants die.

▲ **Figure 2.8** These **maize** plants in south-west France died through lack of water in a drought.

In Chapter 1, you saw how water travels through a plant and escapes through the leaves. Leaves form a large surface area of the plant so this might be the place where the light produces food to keep the plant alive.

Finally, from as far back as Hale's time, scientists knew about carbohydrates. From all this information, we can start to build up an equation for photosynthesis. Van Helmont helps us with the first part of the equation:

water → mass of plant

Testing for the carbohydrate called starch in leaves

Next, we need to find a link to the carbohydrate, and the carbohydrate that is used in investigations on photosynthesis is called **starch**. The test for starch is given below. A different test is used for other carbohydrates.

If iodine solution is added to a sample and starch is present, the brown iodine solution changes the cells of the sample to blue–black.

Starch grains form in the cytoplasm of leaf cells, which also contain chloroplasts. Figure 2.9 shows the distribution of chloroplasts in the cells of a leaf.

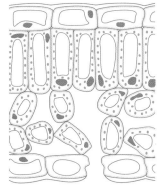

▲ **Figure 2.9** The layers of cells in a leaf. The green dots are chloroplasts.

The chloroplasts are green because they contain the green pigment **chlorophyll**. This colour can prevent you seeing any colour change when testing for starch. The chlorophyll can be removed in the following way:

1 A beaker of water is boiled and the heat source is turned off.
2 The leaf is held in a pair of forceps and dipped in the beaker of hot water for about 30 seconds. This kills the cells and prevents any further reactions from taking place in them. It also makes it easier for the iodine solution to enter the cells in step 5.

Work safely

Take care when working with boiling ethanol – safety glasses are needed to protect your eyes.

Be sure to turn the heat source off, as specified in Step 1 of removing the chlorophyll from a green leaf, as alcohol vapour is very flammable.

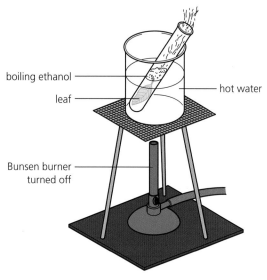

boiling ethanol

leaf

hot water

Bunsen burner turned off

▲ **Figure 2.10** Removing chlorophyll from a leaf.

3 The leaf is then placed in the bottom of a test tube and covered with ethanol. The test tube is placed in the beaker of hot water as Figure 2.10 shows. Ethanol has a boiling point that is lower than the boiling point of hot water and as the ethanol boils, it dissolves most of the chlorophyll in the cells and makes the starch grains easier to see when they are stained with iodine solution, which is used here as a starch indicator. When the iodine solution is introduced to a substance which contains starch, it stains the starch grains blue-black.

The presence of starch grains in a leaf can be revealed in the following way:

4 The ethanol in the test tube is poured into a second beaker and the leaf, which is now brittle, is removed with forceps and dipped into the hot water again to make it softer and easier to handle in the next step.
5 The soft leaf is then spread out on a white tile and drops of iodine solution are released onto its surface. The iodine solution enters the cells and if starch grains are present, they are stained blue–black.

10 Plant cells build up starch and then remove it, as you shall see later. When testing a leaf for starch, what might happen if the cells were not killed in step 2?

11 What would happen if the boiling point of ethanol was higher than that of water?

12 What colour would you expect the ethanol to go after the leaf has been boiled in it? Explain your answer.

13 What might happen to the brittle leaf if you tried to spread it out on the white tile?

14 Why is it important to spread the leaf out on a white tile and not a coloured one?

Removing the carbohydrate called starch from a plant

A plant from which the carbohydrate starch has been removed is called a **de-starched** plant. A plant is de-starched by putting it in a dark place, such as a cupboard, for two or three days. Once a plant is de-starched, it can be used in investigations about photosynthesis.

Which plant has had the carbohydrate starch removed?

Work safely

Take care when working with boiling ethanol – safety glasses are needed to protect your eyes.

Be sure to turn the heat source off, as specified in Step **1** of removing the chlorophyll from a green leaf, as alcohol vapour is very flammable.

Use the information about testing a leaf for starch in the previous paragraphs to help with this enquiry.

Your teacher will provide you with two plants – one that has been placed in a cupboard for two to three days and another that has been kept in full sunlight. You must find out which plant has been de-starched.

Planning and investigating

Write a list of all the equipment you need. Write a step-by-step plan of how you will perform your enquiry. Check this with your teacher and, if approved, try it.

Examining the results

How did the leaves compare?

Conclusion

Which plant had been de-starched?

From this enquiry, we can add to our description (or model) of photosynthesis with this equation:

 water → carbohydrate (when light is present)

15 Does this equation support or refute the one made by van Helmont? Explain your answer.

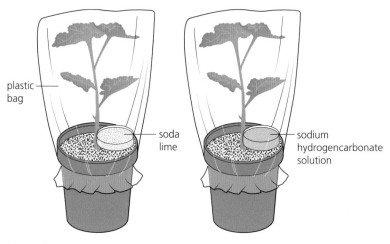

plastic
bag

soda
lime

sodium
hydrogencarbonate
solution

▲ **Figure 2.11** Investigating the effect of carbon dioxide on starch production.

Carbon dioxide and starch production

Returning to Hale's discoveries, we can now test his idea about the importance of carbon dioxide.

To investigate the role of carbon dioxide in starch production, two de-starched plants are needed. One is enclosed in a transparent plastic bag with a dish of soda lime, and the other is enclosed in a transparent plastic bag with a dish of sodium hydrogencarbonate solution. The air in the bag is sealed inside with an elastic band, which holds the bag to the plant pot (Figure 2.11).

Soda lime is a mixture of sodium hydroxide and calcium hydroxide. It has the property of absorbing carbon dioxide from the air. Sodium hydrogencarbonate solution releases carbon dioxide into the air. It enriches the air with carbon dioxide.

The plants are set up in a sunny place and left for a few hours, and then a leaf from each one is tested for starch. A leaf from the plant in the air without carbon dioxide does not contain any starch, but a leaf from the plant in the air enriched with carbon dioxide does contain starch. So we can conclude that carbon dioxide is needed for starch production.

The second step in building the equation for photosynthesis is shown here:

 water + carbon dioxide → carbohydrate (starch)

So far, we have produced a word equation for photosynthesis with two reactants and one product but, before we investigate the final product, we need to look at two other factors involved in photosynthesis and see if we can also fit them into the equation – light and chlorophyll.

16 If you were to test a small sample of soda lime with universal indicator, would it register a pH above or below 7? Explain your answer.

17 What is the purpose of the plastic bag and elastic band?

18 Is it important for the bags to be transparent? Explain your answer.

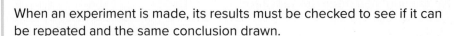

What is the effect of carbon dioxide on starch production?

When an experiment is made, its results must be checked to see if it can be repeated and the same conclusion drawn.

Planning and investigating

Make a list of the items you will need to set up two plants for the investigation. Write a step-by-step plan of how you will set up the plants. If your teacher approves, try it. Leave the plants until the following day.

Make a step-by-step plan to test a leaf from each plant the following day.

Examining the results

Compare the two leaves.

Conclusion

Use your evaluation to draw a conclusion. Does it support the equation below?

water + carbon dioxide → carbohydrate (when light is present)

19 Does this equation support or refute the ones made by van Helmont and Hales? Explain your answer.

Light and starch production

When scientists collect data and draw a conclusion from an experiment, they like to devise another experiment to see if the new data supports or refutes the conclusion of the first experiment. Here, the effect of light on starch production has been confirmed by using a de-starched plant.

Here is another experiment for you to try, to check that light is really necessary for a leaf to produce starch. The experimental plant is set up as shown in Figure 2.12.

One de-starched plant is needed for this investigation to find out if light is needed for starch production. One leaf is covered with aluminium foil, which is an opaque material. Another leaf is covered in transparent plastic.

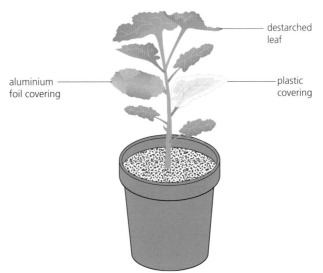

destarched leaf

aluminium foil covering

plastic covering

20 Why do you think a cover was used for the leaf exposed to the light?

▲ **Figure 2.12** Investigating the effect of light on starch production using a de-starched plant.

The plant is left for four hours in a sunny place. Then, the covers are removed from the leaves, and they are tested for the presence of starch. The leaf enclosed in aluminium foil does not contain any starch, while the leaf enclosed in transparent plastic does contain starch. This shows that light is needed for starch production and the next step in building the equation for photosynthesis is:

$$\text{water} + \text{carbon dioxide} \xrightarrow{\text{light}} \text{carbohydrate}$$

You can see that the position of light has been moved out of the equation as it is not a chemical taking part in a reaction.

What is the effect of light on starch production?

Make a list of the items you need to set up the plant as shown in Figure 2.12, as well as those you need in order to test for starch.

Planning and investigating
Make a step-by-step plan for setting up the plant with its covers and for its exposure to light.

When you have collected the leaves, gather the items you need for the starch test and try it.

Examining the results
Compare the two leaves.

Conclusion
Use your evaluation to draw a conclusion. Does it support the equation with light over the arrow?

Chlorophyll and starch production

Most leaves are green. The colour is due to the presence of chlorophyll in the chloroplasts in the leaf. It seems that the chlorophyll could be important in starch production. We know how to take the chlorophyll out of a leaf but when it is done, the leaf is dead. In order to test the role of the chlorophyll in the chloroplasts on photosynthesis, a leaf has to be used which has parts that do not have any chloroplasts. In a variegated leaf (shown in Figure 2.13), the parts without chloroplasts are white.

When a variegated leaf is tested for starch production, it shows that the following equation can be constructed:

$$\text{water} + \text{carbon dioxide} \xrightarrow{\text{light} + \text{chlorophyll}} \text{carbohydrate}$$

Confirm or refute the equation by performing the following enquiry.

▲ **Figure 2.13** This plant has variegated leaves. The white parts show where chlorophyll is absent.

What is the link between chlorophyll and starch production?

Hypothesis

Produce a hypothesis about how a variegated leaf may show that chlorophyll is important for photosynthesis.

Prediction

Make a prediction based on your hypothesis.

Planning and investigating

Make a plan to test your hypothesis. If your teacher approves, try it.

Examining the results

Compare the green and the white parts of the leaf after they have been tested with iodine solution.

Conclusion

Draw a conclusion based on your evaluation.

Science in context

The discovery of oxygen

Joseph Priestley (1733–1804) was an English chemist who studied gases. The equipment used at the time to trap gases was an upside-down container put over the experiment, to catch any gases that were produced. In the course of his investigations, Priestley discovered that sometimes a gas was produced in which things could not burn. When he put a plant, such as mint, in a jar of this gas and let sunlight shine on it, he found that the gas appeared to change to one that did allow things to burn in it. Priestley considered that a gas had been produced which refreshed the air.

Later, Priestley met the French chemist Antoine Lavoisier (1743–94) and told him about his discovery. After thinking about Priestley's investigations and performing some tests on the gas, Lavoisier named it **oxygen**.

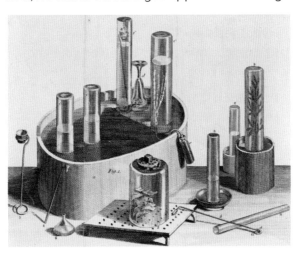

▲ **Figure 2.14** The equipment Joseph Priestley used in his investigations on gases.

Photosynthesis and oxygen

Today the gas produced by plants can be investigated using Canadian pondweed, set up as shown in Figure 2.15. The equipment on the left is placed in a sunny place and the equipment on the right is kept in the dark. After about one week, the amount of gas collected in each test tube is examined. The plants in the dark have not produced any gas but the plants in the light have produced a gas that relights a glowing splint – showing that the gas is oxygen.

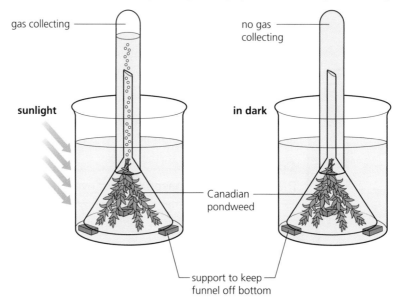

gas collecting

no gas collecting

sunlight

in dark

Canadian pondweed

support to keep funnel off bottom

▲ **Figure 2.15** Equipment for investigating oxygen production.

You can now repeat a version of Priestley's experiment to check his result.

21 Word equations have reactants and products. What are the reactants and what are the products in the word equation for photosynthesis?

DID YOU KNOW?
A few plants have modified their leaves to feed on animals. The Venus fly trap captures insects by snapping parts of its leaves around them, and pitcher plants make long tubes full of fluids that can digest flies, birds and bats!

22 Does your data support the data that Priestley collected over two hundred years ago?

Can you repeat Priestley's experiment?

Hypothesis
Construct a hypothesis based on your knowledge of photosynthesis and Priestley's work.

Prediction
Make a prediction from your hypothesis.

Planning and investigating
Make a step-by-step plan to set up the experiment and carry it out.
Include the test for oxygen at the end of your plan.

Examining the results
Compare the data you collect from the two sets of plants after one week.

Conclusion
Draw a conclusion based on your evaluation of the data.

This information provides the final step in producing the word equation for photosynthesis:

 $$\text{water} + \text{carbon dioxide} \xrightarrow{\text{light} + \text{chlorophyll}} \text{carbohydrate} + \text{oxygen}$$

The perfect environment for growth

When plants are grown for food, every attempt is made to ensure that the crop yield is as large as possible.

Applying **fertilisers** helps crop growth, and the use of **pesticides** keeps other organisms from damaging the crop.

Science in context

Using scientific knowledge to grow more food

In the past, farmers tried to provide the perfect environment for growth in their fields, and in many parts of the world, they continue to do this today. More recently, greenhouses and polytunnels are being used to raise crops for food all over the world. A typical greenhouse has a metal frame which holds transparent sheets of glass. A polytunnel is made out of transparent plastic sheets supported by a metal frame. Both buildings provide a well-lit place in which the temperature and the amount of water, minerals and carbon dioxide in the air can all be controlled.

▲ **Figure 2.16** This passive solar greenhouse, in Hubei Province in central China, is ideal for growing crops, even in cold temperatures. Unlike a typical greenhouse, this structure has a solid wall on the north-facing side (shown on the right in this image), which acts as an insulator. This helps to keep the solar heat inside the greenhouse. Plastic sheeting, rather than glass, is used on the south-facing side of the building, and allows for maximum sunshine to enter the greenhouse during the day.

▲ **Figure 2.17** In this polytunnel, the perfect environment is being created for growing cucumber plants.

Summary

✔ Plants make carbohydrates using energy from light in a process called photosynthesis.
✔ Many plant cells have chloroplasts which contain chlorophyll, and photosynthesis occurs in the chloroplasts.
✔ Photosynthesis can be summarised by this word equation:

$$\text{water} + \text{carbon dioxide} \xrightarrow{\text{light} + \text{chlorophyll}} \text{carbohydrate} + \text{oxygen}$$

✔ Knowledge about photosynthesis developed over time and involved the work of many different people.
✔ An understanding of the process of photosynthesis has been applied to increase crop production.

End of chapter questions

1 Do the word equations for respiration and photosynthesis support each other in some way? Explain your answer.
2 Why must plants be destarched before being used in photosynthesis experiments?
3 A water plant produces a few bubbles of oxygen in a certain time when the temperature of its pond is 10 °C. Predict what will happen to the bubble production when the temperature of the pond water rises to 20 °C. Explain your answer.

 Now you have completed Chapter 2, you may like to try the Chapter 2 online knowledge test if you are using the Boost eBook.

In this chapter you will learn:

- about how individuals within the same species can vary
- about the different types of variation (Science extra)
- about Gregor Mendel, Hugo de Vries and their work on genes (Science in context)
- that chromosomes are found in the nucleus of a cell
- that special cells called gametes combine to produce a fertilised egg
- about the difference between XX and XY chromosomes
- how genes, made of DNA, are responsible for individual variation
- about DNA (Science in context)
- about ideas before the theory of natural selection (Science in context)
- about the theory of natural selection.

Do you remember?

- What do you call animals with a backbone and animals without a backbone?
- How can you tell a fish from a mammal?
- How is an amphibian different from a bird?

Variation in living things

Many living things have certain features in common as well as features which are different.

For example, a cat, a rabbit and a monkey all have ears and a tail. However, these features vary from one kind of animal to the next. For example, in the **species** shown in Figure 3.1 (on the next page), the external parts of the ears of the rabbit are longer than the ears of the cat, and the external parts of the monkey's ears are on the side of its head, while the other two animals have them on the top. The cat and the monkey both have long tails, but this monkey's tail is different because it is prehensile, which means the monkey can wrap it around a branch for support while it hangs from a tree to collect fruit. (Only monkeys that come from Central and South America have prehensile tails.) The rabbit's tail is much shorter than both the cat's tail and the monkey's tail.

By comparing living things, we can classify them into groups. We make and use keys to classify living things in this way. When we work through a classification system, we find that we reach a group called the **species**. You have just seen how different species vary or have different features from each other. Variation also occurs between individuals in the same species.

▲ **Figure 3.1** A cat, a rabbit and a Costa Rican spider monkey have features in common, but they show **variations** too.

Variation within a species

The individuals within a species are not identical. Each one differs from all the others in many small ways. For example, one person may be tall, have small fingers and ears with free lobes while another person may be small, have long fingers and ears with attached lobes. Another person may have different combinations of these features.

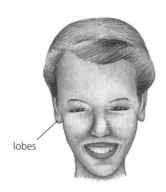

lobes

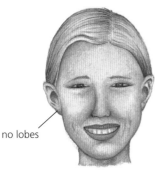

no lobes

▲ **Figure 3.2** Ears with and without ear lobes.

> **Science extra: Types of variation**
>
> There are two kinds of variation that occur in a species. They are called **continuous variation** and **discontinuous variation**:
> - A feature that shows continuous variation may vary in only a small amount from one individual to the next, but when the variations of a number of individuals are compared, they form a wide range. Examples include the range of values seen in heights or body masses of humans.
> - A feature that shows discontinuous variation shows a small number of distinct conditions, such as being male or female, and having ear lobes or not having ear lobes. There is not a range of values between the two, as there is between a short person and a tall person, for example. Another feature showing discontinuous variation is 'hitchhiker's thumb', which is when the thumb in the thumbs-up position bends backwards towards the wrist.

Inherited characteristics

The features of living things, including humans, are also called **characteristics**. When scientists grasped the idea that characteristics could be passed from one generation to the next, the study of **inheritance** began which lead to the discovery of **genes** and the study of genetics.

When you look at a family photograph, you can sometimes see that some members of the family share similar physical features. In other words, those features are found in different generations, which suggests that they could be inherited. You will look at this idea again later in the chapter.

Science in context

Gregor Mendel, Hugo de Vries and genes

Gregor Mendel (1822–84) was an Austrian monk who studied mathematics and natural history. He set up experiments to investigate how features in one generation of pea plants were passed on to the next.

▲ **Figure 3.3** Gregor Mendel.

A pea plant will pollinate itself – it does not need bees to bring **pollen grains** from other flowers, although it can receive them. In most flowers of flowering plants, there is a structure that produces and releases pollen called the **anther** and there is a structure that receives pollen grains called a **stigma**. When a pollen grain lands on a stigma, **pollination** is said to have occurred and then the process of **fertilisation** and seed production follows.

Mendel wanted to control the way the flowers pollinated, so he cut off the anthers of one flower, collected pollen from another flower and brushed it on to the stigma of the first. He completed his task by tying a **muslin** bag around the first flower to prevent any other pollen from reaching it.

Mendel performed thousands of experiments and used his mathematical knowledge to set out his results and to look for patterns in the way that the plant features were inherited. He suggested that each feature was controlled by an inherited factor. He also suggested that each factor had two sets of instructions and that parents pass on one set of instructions each to their offspring. Many years later it was discovered that Mendel's 'factors' were genes.

Although his work was published by a natural history society, its importance was not realised until 16 years after his death. At that time, Hugo de Vries (1848–1935), a Dutch botanist, had been studying how plants pass on their characteristics from generation to generation. He was checking through published reports of experiments when he discovered Mendel's work. He found that his own work supported Mendel's and, after studying Charles Darwin's hypothesis (called pangenesis) to explain inheritance, he suggested that information about

1 Why did Mendel cut out the anthers of some flowers?

2 Why did Mendel tie a muslin bag around the flowers in his experiments?

3 What evidence provided Mendel with the idea of using pollen studies to investigate inheritance?

4 What is the value of performing a large number of experiments?

5 How did Mendel's mathematical knowledge help him in considering the evidence of his investigations?

6 Which scientific enquiry skills was de Vries applying when he discovered Mendel's work?

7 What evidence stimulated de Vries' creative thoughts about inherited particles?

the characteristics to be inherited were in the form of small particles called pangenes. Darwin's hypothesis was shown to be incorrect, but de Vries' idea about pangenes continued to be useful to scientists studying inheritance. A Danish biologist called Wilhelm Johannsen (1857–1927) then made the word shorter and called the inherited particles **genes**.

Chromosomes

As living things are made from cells, scientists began looking at cells for evidence that they may be involved in reproduction. All cells have a cell membrane, cytoplasm and a **nucleus**. When scientists examined living cells under the microscope, they saw that changes took place in the nucleus that are linked to reproduction. Just before a cell divides, long strands of material appear in the nucleus. These strands are called **chromosomes**.

CHALLENGE YOURSELF

Make a model of a cell with chromosomes as shown in Figure 3.4. Select your materials and, if your teacher approves, make your model.

What are the model's strengths and weaknesses? You may need it later for a presentation.

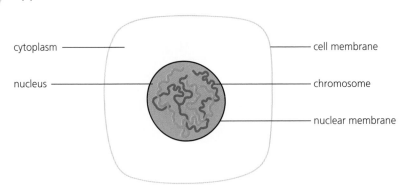

cytoplasm — cell membrane

nucleus — chromosome

— nuclear membrane

▲ **Figure 3.4** Chromosomes in the nucleus of a cell.

Before cell division begins, when the chromosomes are not visible under the microscope, each chromosome makes a copy of itself, and they lie together as two threads. During cell division, as the nucleus divides, each pair of threads separates. Each thread then enters one of the two nuclei that are forming in the new cells. When each cell is formed, every thread becomes a chromosome. This means that the nuclei of the new cells have the same number of chromosomes as the original one. The chromosomes in the nuclei of the new cells then seem to disappear into the nuclear material but become visible under the microscope once more when the cells are about to divide.

DID YOU KNOW?

Chickens have 78 chromosomes while elephants have 56, a starfish has 36, rice has 24 and kangaroos have 16!

Extensive studies on chromosomes have shown that they are arranged in pairs in the nucleus of the body cells of living things, and that certain individuals in each species have a certain number of chromosomes in their cell nuclei. For example, humans have 46 chromosomes arranged in 23 pairs in every nucleus, and a fruit fly has eight chromosomes arranged in four pairs.

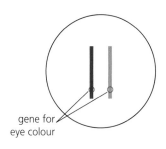

gene for
eye colour

▲ **Figure 3.5** Two genes on a pair of chromosomes.

Chromosomes and genes

Chromosomes are like threads of chemical messages. These messages are strung along each chromosome like the carriages of a very long train. Each message is called a gene. The genes provide all the information for how the cell grows, develops and behaves and how the body grows and develops too. Each pair of chromosomes has pairs of genes that carry information for a particular characteristic, such as eye colour or hair colour. These genes are situated at the same point on each chromosome, as Figure 3.5 shows.

During the formation of **gametes**, parts of the chromosomes swap portions. This swapping leads to a mixing up of the genes, so an exact copy of the parent's genetic code is not passed on. Each gamete only has half the number of chromosomes as the other cells in the body. However, after fertilisation, the new nucleus contains all the chromosomes and genes needed to make the new individual. As there has been some mixing of the genes from both parents, the new individual develops a slightly different combination of features from their parents, which leads to variation in the species.

Figure 3.6 shows a family. The father has genes for black hair, curly hair, blue eyes, attached ear lobes (where the lower part of the ear is joined straight down on to the side of the head) and for the absence of freckles. The mother has genes for red hair, straight hair, brown eyes, free ear lobes and freckles.

Father   Mother   Alberto

Benita   Carlos   Dorita

▲ **Figure 3.6** Variation in a family.

8 What features have
  a Alberto
  b Benita
  c Carlos
  d Dorita

inherited from their mother and from their father?

You can see that the characteristics controlled by genes can vary between one generation and the next (parents and children) and among individuals of the same generation (the children).

Sex chromosomes

When scientists discovered how to take photographs of chromosomes in a cell, they cut up the photographs and arranged the chromosomes into pairs. When they did this with cells from a male, they found that one pair appeared different to the others – one chromosome was larger than the other. They called the larger chromosome the X chromosome and the smaller one the Y chromosome. When they examined the cells from a female, they found that there were two X chromosomes present.

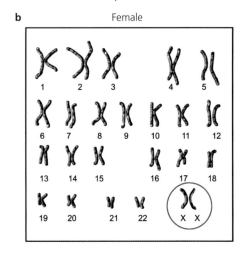

a Male   **b** Female

▶ **Figure 3.7** Pairs of chromosomes from **a** a male and **b** a female.

From their discovery, they worked out how the sex of a person was inherited. They used a genetic diagram called a **punnet square** as shown in Table 3.1. The punnet square shows the possible combinations of the sex chromosomes at fertilisation. The two X chromosomes that the female can produce are shown at the top of the table. The X or Y chromosomes that the male can produce are shown in the left-hand side of the table. The four different combinations are shown within the table. This reveals that there is a 50 per cent chance of an X chromosome combining with a Y chromosome.

9 Why does the combination of the sex chromosomes lead to the chance that equal numbers of boys and girls will be born in a **population**?

▼ **Table 3.1** A punnet square showing the possible sex chromosome combinations.

		Female chromosomes in a gamete	
		X	**X**
Male chromosomes in a gamete	**X**	XX	XX
	Y	XY	XY

Chromosomes and gametes

Once the link between chromosomes and the reproduction of body cells had been worked out, scientists looked for a link between chromosomes and the cells involved in the reproduction of whole individuals. These cells are called the sex cells or gametes. In animals, the male gamete is the sperm and the female gamete is the egg or ovum. In plants, the male gamete is a cell enclosed in a pollen grain and the female gamete, called the egg cell, is in the ovule. During fertilisation, the nuclei of the male and female gametes join together, as Figure 3.8 shows. This fusion of the gametes produces a fertilised egg which then grows into a new individual.

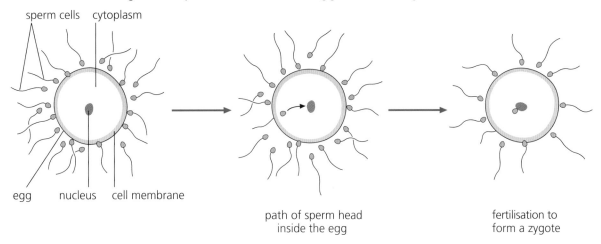

path of sperm head
inside the egg

fertilisation to
form a zygote

▲ **Figure 3.8** Fertilisation.

If the gametes were produced in the same way as ordinary body cells, they would have the same number of chromosomes as body cells. This would create a problem at fertilisation as the fertilised egg would have twice as many chromosomes as the parents. If the fertilised egg from this generation formed individuals that bred, the cells of the new individuals would have four times as many chromosomes as the grandparents. After a few more generations, the cells would be so packed with chromosomes that they would die. This does not happen because there is a special type of cell division that takes place in the reproductive organs, which produces gametes with half the number of chromosomes as the other body cells. In this type of cell division, the chromosomes do not make copies of themselves but instead the pairs separate and move into the newly forming cells – the gametes.

For example, the cells in the human reproductive organs pass on 23 chromosomes into each of the gametes they produce. At fertilisation, the chromosomes from the gametes pair up in the **zygote** (a fertilised egg) and the production of body cells begins again, as Figure 3.9 on the next page shows.

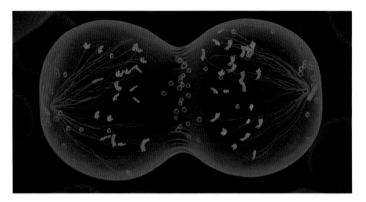

▲ **Figure 3.9** The production of body cells at fertilisation.

10 The cells of a pineapple plant have 50 chromosomes in their nuclei.
  a How many chromosomes are there in the male nucleus in a pineapple pollen grain?
  b How many are there in the female nucleus in the ovule of a pineapple flower?
  c How many chromosomes are in the cells of a seedling?

DNA

Genes are made from a substance called **deoxyribonucleic acid**, which is usually shortened to **DNA**. Finding the structure of DNA and how it is used in genes and chromosomes took many years and involved the work of a wide range of different scientists.

As each person's DNA is unique, it can be used for identification purposes. A person's DNA profile (sometimes called a DNA fingerprint) can be made from cells in the saliva or the blood. The DNA is chopped up by **enzymes** and its pieces are separated into a gel in a process like chromatography. (Remember that chromatography is the process used to separate colours in an ink by putting a drop of ink onto a paper and allowing water to soak up through it.) The pattern of the pieces looks like a barcode on an item of goods. Closely related people have more similar profiles than those who are not related.

DNA and fertilisation

We know that chromosomes are threads of chemical messages, which we call genes, and that genes are made from a substance called DNA. We also know that plants and animals produce sex cells, called gametes, which contain chromosomes and genes. During gamete formation, parts of the chromosomes swap portions. This means that they swap genes and, as we now know, this means that they swap portions of DNA. This in turn means that the gametes which are passed to the next generation have a slightly different combination of DNA from the parent generation.

In the process of fertilisation, a male and female gamete fuse together. The DNA from the two parents forms the chromosomes of the new cell, called a zygote. The combination of DNA molecules in the new cell is slightly different from that of either parent, which leads to the offspring having slightly different features from both of them.

Science in context

Finding out about DNA

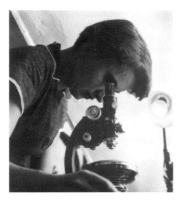

▲ **Figure 3.10** Rosalind Franklin's evidence was vital in the discovery of the structure of DNA.

The first work on investigating chemicals in cell nuclei was carried out in 1869 by Johann Friedrich Miescher (1844–95). He used the white cells in pus and the substance he discovered was called nuclein. Over the next 84 years, generations of scientists made further investigations on this substance. Rosalind Franklin (1920–58) studied the structure of molecules by firing x-rays at them. In 1951, she investigated DNA in this way and her results suggested that it could be made of two coiled strands, but she was not sure.

In 1953 James Watson and Francis Crick, using some of Franklin's results to help them, worked out that DNA is made from long strands of chemicals that are coiled together to make a structure called a **double helix**. The chemicals are arranged in a sequence that acts as a code. The code provides the cell with instructions on how to make the other chemicals that it needs to stay alive and develop properly.

▲ **Figure 3.11** James Watson and Francis Crick used evidence from several sources to work out the structure of the DNA molecule.

Barbara McClintock (1902–92) was a biologist who studied maize. While she was still a student, she worked out a way of relating the different chromosomes in the nucleus to the features of the plant. Later, in the 1940s, she discovered that the genes on a chromosome could change position. They became known as 'jumping genes'. This discovery did not fit in with the way genes were thought to act and her work was not accepted by other scientists. But in the 1970s, during investigations by scientists on the DNA molecule, it was found that parts of the DNA broke off and moved to other parts of the chromosome. McClintock's work was proved to be

LET'S TALK

In your opinion, how firmly should scientists hold their views? Share your ideas with a partner.

11 What preliminary work did Rosalind Franklin do before she investigated DNA?

12 How did Franklin's work help James Watson and Francis Crick?

CHALLENGE YOURSELF

Look at Figure 3.12 to help you make your own model of a section of DNA. Select your materials and prepare a plan. If your teacher approves, try it.

CHALLENGE YOURSELF

Groupwork

Use the models you have made in the last two challenges to give a presentation about chromosomes and DNA. How will you use them and what will you say? Work out a plan as a group, then film your presentation.

LET'S TALK

How could DNA be used to investigate a crime?

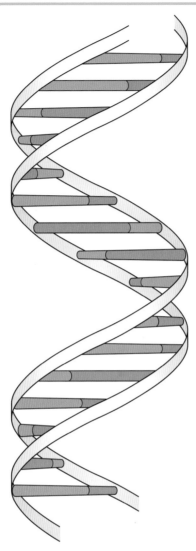

▲ **Figure 3.12** The basic structure of DNA.

correct and in 1983 she received the Nobel Prize for Physiology or Medicine.

In 1990, the Human Genome Project was begun in order to identify up to 25 genes on human chromosomes. This was an international project with scientists from many countries around the world working together to make it possible to see all the genes that are required to form the human body. The project was completed in 2003 and the data it has produced is being used in investigations in medical research to cure certain diseases, in the study of micro-organisms that cause diseases and decompose wastes, in the study of chemicals and radiation that is harmful to us and in the study of forensic science.

▲ **Figure 3.13** Li Ho (left) was a Chinese-American High School student who worked on the Human Genome project with Dr Greg Lennin, a senior biomedical researcher.

◀ **Figure 3.14** Barbara McClintock with the trophy for the Albert Lasker Award for Basic Medical Research she won in 1981.

Genetic vocabulary

When geneticists (scientists who study genetics) speak or write about their work, they often link the terms used within a sentence, so it is important to be clear about meanings and how terms link together.

▼ **Table 3.2** Important genetic terms.

Term	Definition
Chromosome	A thread-like structure that appears in the nucleus of a cell when the nucleus divides. It is made of DNA.
Gene	A section of a chromosome that contains the DNA that holds information about how a particular characteristic, such as hair colour or eye colour, can develop in the organism.
DNA (deoxyribonucleic acid)	A molecule which has two strands of smaller molecules coiled around each other. The arrangement of the smaller molecules form packages of information, called genes, which control the development of an organism.
Genetics	The study of genes and how they pass from one generation to the next and produce variation between individuals.

The scientific theory of natural selection

Scientific enquiry begins with a hypothesis that can be tested. A hypothesis is a suggestion of how something seems to be. A prediction is made that is based on the hypothesis, then an experiment or investigation is carried out, data is collected and compared with the hypothesis and prediction. It does not matter if the evidence provided by the data supports or refutes the hypothesis – either way, a scientific fact is discovered. More experiments and investigations are made to test the hypothesis and eventually enough facts are discovered in order to be certain that the hypothesis, or a modification of it, is a true explanation of the facts. When this stage is reached, the hypothesis or its modification becomes a theory. The theory of **natural selection** is an example. It was set up or formulated by Charles Darwin and is used to explain how one species can change into another over a long period of time.

> **Science in context**
>
> **Ideas before the theory of natural selection**
> Some scientists have been fascinated by their observations on different species and have tried to discover a relationship between them. When Darwin began his studies, most scientists believed that the Earth had existed for only a few thousand years. However, the work of scientists called geologists, who study rocks, were providing evidence that the Earth was much older.

Most scientists believed in the ideas of the Ancient Greek philosopher Aristotle (384–322 BCE). Aristotle had looked at the living things around him and seen that some had simpler bodies than others. This led him to believe that living things could be grouped and placed on a ladder, called the ladder of nature, with the simplest on the bottom rung and the most complex (humans) on the top rung. Aristotle also believed that all species had existed since the beginning of life on Earth and had remained unchanged throughout history.

▲ **Figure 3.15** This picture of Comte de Buffon shows him with some of the living things he studied.

One scientist who doubted Aristotle's ideas was Comte de Buffon (1707–88), a French naturalist. From his studies on plants and animals from around the world, he believed that in the past, plants and animals had developed in one place and then spread out and responded to the different environments by making changes to their bodies. He could not explain how the changes might have come about.

Jean-Baptiste Lamarck (1744–1829), another French naturalist, also believed that one species developed into another and then another as conditions changed. He did not believe that a species could become extinct. He believed that the way a species changed was by developing a feature and then passing it on to future generations. At the time, the giraffe had been newly discovered by Europeans and he used it to explain his ideas.

▲ **Figure 3.16** A giraffe feeding in the treetops.

He thought that it had developed from a smaller antelope that stretched its neck, legs and tongue to feed in the treetops. These stretched features were then passed onto the offspring and when they stretched to reach higher, their necks, legs and tongue became even longer. This theory was known as the theory of acquired characteristics.

LET'S TALK
Giraffes have dark areas on their coat to help them hide from **predators**. Can Lamarck's theory of acquired characteristics explain how the giraffe got them? Explain your answer.

Darwin and his work

English naturalist Charles Darwin (1809–82) was employed on a ship called the *Beagle*, which was about to make a journey around the southern hemisphere in 1831. The trip was originally planned to take two years, but instead took five. The *Beagle* left England and sailed to South America, then into the Pacific and on to the Galapagos islands, before sailing to Australia, then across the Indian Ocean to the tip of South Africa, then back to South America before returning home.

Darwin was mainly concerned with geological specimens on the trip, but when he was on the coast of Chile, he observed mockingbirds. Later, he visited the Galapagos islands, just over a thousand kilometres west of the Chilean coast, and discovered more mockingbirds. He found that they all differed from those in Chile and he also discovered that the mockingbirds on each island differed from each other. These observations made him seek an explanation and he thought that the species on the coast reached the Galapagos islands, but as it settled on each island, it developed features to help it survive.

Darwin was also interested in birds called finches, which he found on the Galapagos islands and in Chile as well. He examined the specimens of finches collected and discovered that they too showed variations between islands, particularly in the size and shape of their beaks, and they also resembled a species of finch found in Chile. He believed that they could have developed in the same way as the mockingbirds, but with these specimens, the beaks gave him an idea about how natural selection might take place. He reasoned that on each island, the birds adapted to feeding on a particular food. **Adaptation** and natural selection became linked in his mind. He set out his thoughts as follows:

- Animals and plants produce a large number of offspring. This number is far higher than the number of parents. For example, a fly lays a large number of eggs and a plant such as the dandelion releases many wind-dispersed fruits.
- The size of a plant or an animal population in a **habitat** usually stays the same, but a change in the habitat (such as the removal of plants and exposing the soil) may cause a change in the populations of plants as they grow back onto the exposed patch.
- If a living thing produces a large number of offspring yet only a few survive, there must be a struggle for survival, and this could be due to a limited supply of food or some other feature of the habitat.
- When you look at a population of any living organism, you can see that the individuals vary from each other.

▲ **Figure 3.17** Charles Darwin.

▲ **Figure 3.18** Mockingbirds on a beach of one of the Galapagos Islands.

13 What creative thought did Darwin have about the mockingbirds and finches on the Galapagos?

14 What two pieces of evidence did Darwin consider when deciding that there must be a struggle for survival among the individuals in a species?

15 What piece of evidence did Darwin consider when deciding that new species are produced by natural selection?

In any species there is competition between the variety of individuals in the population. Some individuals in a species have features that are well adapted to their conditions. They have a better chance of survival than individuals with less suitable features. As a consequence, the best-suited individuals will leave more offspring than the others and these may also have the favourable features. If this continues for some time, a new species may develop. This is the process that Darwin called natural selection. Natural selection gives us an explanation as to how the different species of finch and mockingbird evolved on the Galapagos islands.

Using the theory of natural selection, the finches that Darwin collected can be set out to show how they all evolved from one type of finch, called the **common ancestor**.

In Figure 3.19, you can see that the ground finch has a blunt beak and feeds on seeds, and the warbler-like finch has a pointed beak and is an insectivore.

This observation suggests the hypothesis that the shape of the beak may help the bird to pick up its food. While the observation of the birds may show the birds eating their foods, an investigation using model beaks picking up the same food may help to confirm or refute the hypothesis.

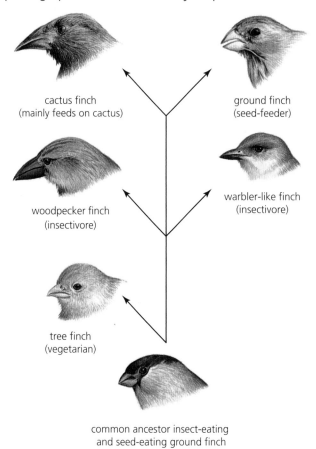

cactus finch
(mainly feeds on cactus)

ground finch
(seed-feeder)

woodpecker finch
(insectivore)

warbler-like finch
(insectivore)

tree finch
(vegetarian)

common ancestor insect-eating
and seed-eating ground finch

▶ **Figure 3.19** Darwin's finches.

16 What do you think an insectivore eats?

Why do birds' beak shapes vary?

You will need:

two similar-sized dishes (for example, petri dishes), a pair of fine forceps (to model a pointed insectivore beak) and a pair of blunt forceps (to model a finch beak for eating seeds), cress seeds (food for the finch), a stop-clock or timer, a partner.

Hypothesis

The shape of the beak can help a bird feed on a certain food.

Prediction

Make your own prediction.

Process

1 Pour about a hundred cress seeds into one of the dishes and spread them out.
2 Set up the second dish close to the first. This second dish is a model of a bird's stomach.
3 Start the stop-clock or timer, and let your partner use the blunt forceps to peck at the seeds, pick them up and transfer them to the second dish (the bird's stomach). This can be done many times, just as a bird naturally pecks at food.
4 Stop the stop-clock or timer after one minute, and count up the seeds in the 'stomach' and record the data in a table.
5 Repeat steps 1–4 with the fine forceps.

Examining the results

Compare the results of the two activities.

Conclusion

Compare the evaluation with the hypothesis and the prediction. Does the data support or refute the hypothesis and prediction?

What are the limitations of this experiment? How could you improve the experiment to make the data more reliable?

In groups, carry out a second investigation based on your answers to the questions in the two paragraphs above. Share the results of your modified experiment with the other groups in your class. Can further improvements be made? Explain your answer.

Natural selection and genetic changes

Darwin didn't know about chromosomes, genes and DNA. All that knowledge came much later, but it supports his theory. We can think of natural selection and genes in the following way:

- An organism has a certain combination of genes. They help the body develop so that it can survive in its environment.
- If the environment changes, the combination of genes will threaten its survival. If one of the genes produces a feature which is particularly suited to the change in the environment, it will help the organism survive. This will increase the organism's chances of breeding and then passing on the beneficial gene. If this happens, more offspring are produced with this beneficial gene, and they survive too.
- Some of the offspring may also have genes that are favourable to the change in the environment, and these are also passed on to future generations. In time, there may be organisms with such a different combination of genes to the original parent population that they form a new species.

LET'S TALK

Imagine there was an island where there were populations of the two types of bird in the enquiry and only one type of food, such as seeds. Now imagine that predators such as hawks arrive on the island and they can catch and feed on both types of bird.

1   How does the presence of the predator affect the time that the birds have to feed? Explain your answer.
2   Will both types of birds be well-fed with the predator on the island? Explain your answer.
3   Will the presence of the predator on the island affect the populations of the two types of bird? Explain your answer.

Summary

- ✔ There is variation between individuals within a species.
- ✔ The nucleus of a cell contains chromosomes.
- ✔ The characteristics that living things possess are inherited.
- ✔ The inheritance of sex is determined by the combination of the X and Y chromosomes – females are XX and males are XY.
- ✔ Chromosomes separate in gamete formation and the gametes fuse to form a fertilised egg.
- ✔ Genes are located in chromosomes and are made from DNA.
- ✔ Knowledge about the role of genes in inherited characteristics and the structure of DNA involved a wide range of people, working over a long period of time.
- ✔ Our current knowledge about genetics can be used to evaluate some early ideas about how characteristics are inherited.
- ✔ Living things evolve due to natural selection.

End of chapter questions

1 What does the word 'inherited' mean?
2 What is a characteristic?
3 How may a litter of kittens show variation?

▲ **Figure 3.20**

4 What does the word 'vary' mean?
5 What is the difference between a chromosome and gene?
6 Why is DNA important in explaining how living things inherit characteristics from their parents?
7 A flock of birds with beaks 6 cm long settle on an island. They use their beaks to search deep within the sandy shore for worms and molluscs to eat. These invertebrates live 5 cm below the surface of the sand. The birds breed and produce offspring with beak lengths from 4–9 cm long.
  a As the offspring grow up and begin to feed on the shore, what do you think will happen to them? Explain your answer.
  b In time, the worms and molluscs change their behaviour and burrow down to 8 cm to live. Many years later, a scientist is the first to visit the island since the original flock of birds arrived. She knew from a previous scientist's notes about the characteristics of the birds. Why may she be surprised at how she finds the birds now? Explain your answer.

 Now you have completed Chapter 3, you may like to try the Chapter 3 online knowledge test if you are using the Boost eBook.

In this chapter you will learn:
- how a developing baby depends on its mother during pregnancy
- how the mother's health and fetal development can be affected by diet, smoking and drugs
- about diseases that can affect a developing fetus (Science extra)
- about neonatal care and the development of incubators (Science in context).

Do you remember?

- What does the term 'reproduction' mean?
- What do the following terms mean in plant reproduction?
  - pollination
  - fruit
  - seed
  - germination
- What is puberty?

▲ **Figure 4.1** A mother goat with her young and bird chicks being fed in their nest. We are familiar with mammals and birds caring for their young after they are born or hatched, but care starts before that as this chapter on human fetal development shows.

Care during fetal development

▲ **Figure 4.2** A mother during pregnancy.

When humans reproduce, the mother carries the developing baby in an organ called the **uterus** for about nine months. During this time, the mother is said to be **pregnant** and the nine-month period is known as **pregnancy**. A developing baby is called a **fetus**, and it receives all its nutrients and oxygen from the mother through a structure called the **placenta**, which is attached to both the mother and the baby. Expectant mothers must be careful about what they take into their body, as these substances can be passed on to the fetus.

Diet

During pregnancy, the mother should eat a balanced diet in order to receive all the nutrients she needs for her own health and the health and growth of the fetus. In places where poverty prevents people from having a balanced diet, the mother's health may suffer and, as a result, the growth of the fetus slows down. When the baby is born, it is smaller than normal and less able to resist the attack of diseases.

In places where there is a greater variety of food in the mother's diet, soft cheeses such as camembert should not be eaten, as they can contain micro-organisms that cause listeria, which can be fatal to a developing fetus. Also, caffeine – a substance found in coffee, chocolate and tea – must be consumed in smaller amounts than usual as it can slow down the growth and development of the fetus.

> **LET'S TALK**
>
> What could be done to help mothers around the world who do not have a balanced diet? How can you find out where they are? How could help be provided?

▲ **Figure 4.3** Everyday foods like camembert and coffee can be dangerous to a developing fetus.

Smoking

When a person smokes, they inhale nicotine and other poisons from the burning tobacco which pass into the bloodstream. If a mother smokes during pregnancy, these substances travel in the blood and are passed on to the developing fetus.

Substances from tobacco lower the amount of oxygen that blood can carry. This means that the fetus has a reduced supply of oxygen needed for survival and healthy growth. The fetus responds by increasing the rate at which its heart normally beats to try and get more oxygen to its cells.

Smoking in pregnancy also increases the risk of the baby being born prematurely (before the end of the usual nine-month period) and having a smaller birth **weight**. It may also cause damage to the lungs and the nervous system and can even result in a stillbirth, which is where the fetus dies before it is born.

> **LET'S TALK**
>
> You are going to produce a poster or handout about smoking and pregnancy for expectant mothers.
>
> How will you present it – as text, or text and illustrations? What will the illustrations be? Do you think it should include scientific details, perhaps in a table, or in an illustration showing a fetus and the places that could be damaged? Will the tone of your poster or handout be a gentle warning, or will it be stronger to act as a shock for pregnant women who smoke?
>
> Make your poster or handout, then work in a group to share and compare your ideas. What is similar and what is different about the posters?

Drugs

Alcohol is classed as a drug because it affects the nervous system and slows down many actions of the body. Non-medicinal (or 'recreational') drugs are drugs which are not used to treat disease or help restore a body to health. These non-medicinal drugs include cannabis, cocaine, heroin and ecstasy. People become addicted to alcohol and non-medicinal drugs for a variety of reasons, and they need special care at health centres and rehabilitation centres to break their addictive habits and become healthy again.

The table shows how consuming alcohol and non-medicinal drugs during pregnancy can affect the fetus.

▼ **Table 4.1** The effects of alcohol and drug use during pregnancy.

Drug	Effect on fetus
Alcohol	<ul><li>Drinking even small amounts can cause nerve damage.</li><li>Drinking large amounts can cause the head to be smaller and cause abnormal development of the eyes, nose and lips.</li><li>Fetal alcohol syndrome causes the effects listed above, as well as further issues after the baby is born, including poor co-ordination and memory, inability to concentrate well and being more active than normal (hyperactivity).</li></ul>
Non-medicinal drugs	<ul><li>These slow down fetal growth.</li><li>They make a fetus less able to fight disease.</li><li>The baby can be born dependent on the drug that the mother was taking.</li></ul>

LET'S TALK

Imagine that you were tasked with communicating the dangers of non-medicinal drugs to developing babies to women at a rehabilitation centre. What would you produce to gently inform them of the dangers? Is your approach the same as for dealing with smoking? Explain your ideas.

CHALLENGE YOURSELF

Imagine you have a relative who is expecting a baby. They want to find out more about how to care for the baby as it develops, is born and is in its early stages of life. Would you advise them to:

- ask other relatives for advice
- seek out health advice from a clinic, nurse or doctor
- search the internet for advice
- look in magazines for advice?

Do some research and consider which sources of information will be more reliable than others. Decide whether they are very reliable, reliable, or not very reliable. Write a conclusion to explain why you think your relative should use some sources for advice and perhaps not use others.

Science extra: Effects of diseases on the fetus

Diseases that affect adults and children can also affect a developing fetus. If a mother has malaria, for example, it can potentially cause the fetus to die, or may slow down its development so that a smaller baby than normal is born. Tuberculosis during pregnancy may also cause the death of a fetus.

Some diseases can damage the fetus, which results in the baby being born with a disability of some kind. For example, rubella (which is also known as German measles, because German doctors first identified it 250 years ago), is a disease which is transmitted by droplets which are coughed or sneezed into the air. It can damage the nervous system of the fetus and result in deafness and blindness. The heart can also be damaged.

If a mother has a sexually transmitted disease (STD) such as chlamydia, gonorrhoea or syphilis during pregnancy, it can cause blindness, deafness, a swollen liver and skin sores in her baby. A mother with AIDS may pass on the virus during pregnancy or at birth and this will damage the immune system of the baby.

Science in context

Neonatal care

The term **neonatal** refers to the period of time from birth until 28 days later. If a baby is ill or is born with some of the conditions mentioned in this chapter, it will need extra care to survive during this time, which can be provided by placing the baby in an incubator.

A major problem for a baby that is born early can be controlling and maintaining its body temperature at a healthy level. If the baby cannot do this, as its temperature drops, **hypothermia** can set in and the baby may die. In the nineteenth century, babies in danger of developing hypothermia were placed in baskets with blankets and hot water bottles. Despite this care, many babies continued to die.

In the 1870s, the French doctor Stéphane Tarnier began to search for a way to solve the problem. When Tarnier visited the Paris zoo, he saw a chicken incubator being used to hatch eggs and protect the young chicks. This encouraged him to invent a similar piece of equipment for human babies. His first invention had enough room for several babies to be warmed together, just like chicks in their incubator. He later modified his invention so that it only held one baby in order to provide greater individual care.

In 1889, another French doctor, Alexandre Lion, modified the basic design of the incubator by adding a thermostat to control the temperature and a fan to keep a flow of air moving over the baby to provide it with oxygen. Later, during the twentieth century, further developments were made which resulted in the creation of the incubators that hospitals use today.

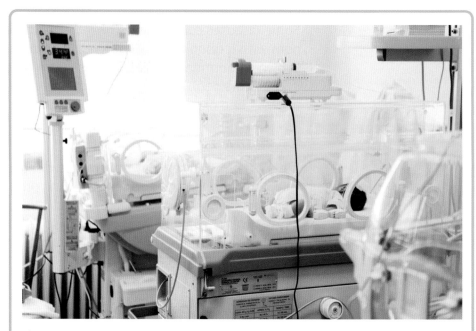

▲ **Figure 4.4** Babies in incubators at a hospital.

CHALLENGE YOURSELF

Make a model incubator and explain what each part does to help a newborn baby survive.

An incubator may be a closed transparent box with openings in the side, or it may have an open top. It may have a number of features, which include a temperature and **humidity** regulator to keep the baby warm and to provide air at the correct moisture level to prevent dehydration. It might have an extra oxygen supply, a ventilator to blow air into the lungs, feeding tubes and built-in life-sign monitors for checking the temperature, breathing and heart rate of the baby inside. In time, the baby may be able to leave the incubator and join its family when it is strong enough.

1 What was the issue that required scientific understanding to help babies survive?

2 How was scientific knowledge applied to the issue?

3 What effect do you think the use of incubators has had on societies around the world?

▲ **Figure 4.5** A newborn baby returned to its family.

Summary

✔ A balanced diet is needed during pregnancy for the healthy development of the fetus.
✔ Smoking can negatively affect the growth and development of the fetus.
✔ Drugs can harm the fetus in a variety of ways.
✔ The development of incubators to support babies that need extra care is an example of how scientific knowledge can be applied in society.

End of chapter questions

1 Imagine you were part of a health team sent into an area which had suffered famine. How would you expect to find the condition of newborn babies?

2 Two mothers are pregnant. One smokes and one does not.
   a Are there any differences between the developing fetuses? Explain your answer.
   b Will both babies be equally healthy when they are born? Explain your answer.

3 Predict the condition of a baby born to a mother who:
   a regularly drank large amounts of alcohol through the pregnancy
   b took a range of non-medicinal drugs through the pregnancy.

Now you have completed Chapter 4, you may like to try the Chapter 4 online knowledge test if you are using the Boost eBook.

Environmental change and extinction

In this chapter you will learn:

- how organisms adapt to the environment they live in
- about observations of chimpanzees (Science in context)
- how environmental changes can affect the populations of organisms that live there
- how a species can become extinct as a result of environmental change
- how humans can cause environmental change
- how an accidental discovery saved a snail species from extinction (Science extra)
- about global extinction events on Earth.

Do you remember?

- What do ecologists mean by the following terms?
  - habitat
  - bioaccumulation
  - invasive species
- What is a fossil?
- Name the three different ecosystems shown in Figure 5.1.
- What is a habitat survey?
- Look at the equipment that students are using in Figure 5.2 on the next page. Name each item and describe what it is for.

▲ **Figure 5.1** Ecosystems of the world.

▲ **Figure 5.2** Equipment used in a habitat survey.

▲ **Figure 5.3** Elephants roaming in South Africa.

The environment

Environmental change and extinction are related, but in order to understand this relationship, we need to review what we already know about the environments, how we survey different environments and how we can study the organisms that live in them.

When you go to the countryside and look at the environment around you, you will see many plants, and animals such as birds and insects flying around. You may also see rocks and soil. You can look up at the sky and make a note of the weather conditions. Look at this natural environment in South Africa.

1 Describe the environment shown in Figure 5.3.

How does a habitat change over time?

You will need:
An area in the school grounds about a square metre in area, which has been cleared of all plants and just has a surface of topsoil.

Hypothesis
Construct a testable hypothesis about whether you think the new-made habitat will change over time. Think about what might be in the soil and the living things in the surrounding habitat. Explain your thinking for the hypothesis.

Prediction
Make a prediction about how the habitat might change, if you believe it will change.

CHALLENGE YOURSELF
Describe the environment around your school at a certain time of day. Can you see features such as plants, animals, rocks or soil? What is the weather like? Explain your answer.

Planning, investigating and recording data

How will you survey the habitat to check your hypothesis and prediction? What data, if any, will you record, and for how long will you collect data? How will you record it?

Examining the results

Examine your data and identify any trends or patterns. Do all results fit the trends?

Conclusion

Compare your evaluations with your hypothesis and prediction and draw a conclusion.

Organisms in the environment

When we adapt to something, we adjust to it. For example, if you change schools, you may find that the new school runs in a slightly different way to the old one and, in time, you adjust or adapt to it. In a similar way, plant and animal species are adapted to their habitats. The features they have that help them survive there are called **adaptations**.

These adaptations have developed through natural selection, as we saw in the last chapter, and mean that a species is 'locked in' to the conditions in the habitat. All living things are adapted to their environment. The plant in Figure 5.4, called a moss campion, is adapted to the mountain environment by growing low to the ground and forming a cushion shape to avoid the strong, cold winds rushing across the mountain peaks.

2 What do you understand by the term 'locked in' in the context of habitat studies?

3 If the average temperature of the climate in the mountains increases, do you expect the population of moss campion plants on the mountain top to change? Explain your answer.

▲ **Figure 5.4** Moss campion can grow very low on the ground on alpine surfaces. The flowers are produced on the side of the moss cushion that receives that most light.

Science in context

Observing chimpanzees

Louis Leakey, a Kenyan archaeologist, put forward the idea that as chimpanzees are closely related to humans, their behaviour may provide some information about early human behaviour. He asked Jane Goodall, who had a great knowledge and enthusiasm for animals, to observe a group of chimpanzees closely and note their behaviour. She originally worked alone, but was later joined by colleagues from around the world. Their research produced a huge amount of information about the lives of chimpanzees and how they are adapted to their environment.

A few examples of her observations include the use of opposable big toes (like human opposable thumbs working against the hand and fingers) to grip branches as the chimpanzees swing through the trees. It was previously thought that chimpanzees were mostly **herbivores** and, although they had large canine teeth, it was thought that they were used for display. Goodall observed the chimpanzees killing and eating mammals such as colobus monkeys. It was also believed that humans were the only animal species to use tools. (A tool is a device that is used to make a task easier.) One day, Goodall observed a chimpanzee poking a grass stalk into a termite mound and raising it to his mouth. When he had gone away, she did the same and found that when she pulled out the stalks, the termites were biting them so hard that they stuck to it. The grass stalk was being used as a tool to obtain food. Shortly after this observation, Goodall saw the same chimpanzee take twigs, strip off their leaves and use them like the grass stalk to feed on termites. She saw other chimpanzees also do this. They were not only using tools but making them too.

▲ **Figure 5.5** Jane Goodall.

4   What qualities did Goodall possess that Leakey thought might be useful in his research?

5   What new data did Goodall discover about how chimpanzees move through the trees?

6   What new knowledge did Goodall discover about the canine teeth of chimpanzees?

7   What new knowledge did Goodall discover about how chimpanzees feed?

CHALLENGE YOURSELF

Groupwork

If an ecologist spends a large amount of time in the environment studying one species, a great deal can be found out about the way of life of a population, as Jane Goodall found out.

Use the internet to find out about the Jane Goodall Institute. Look at the section on roots and shoots, and discuss with your group the possibility of setting up a project. Then discuss your ideas with your teacher to see if you can take them forward.

Ecological models: The food web

By making surveys and studying living things in their environments, ecological models can be made. One of the first ecological models you made was the **food chain**.

By looking at feeding relationships between living organisms in an environment, food chains can be built up and linked together to make a **food web**, which provides a model of the feeding relationships of all the living things in the environment. Figure 5.6 shows an example of a food web for a European woodland. Careful observation of the producers has allowed specific plant parts to be used as the basis of the food chains. It shows the movement of food through the environment (the woodland). Food webs can also be used to predict what might happen to populations in a habitat if one of the links in the food web were absent.

8 How might the populations in the woodland change if the following were to happen?
   a The population of foxes were removed.
   b All the seeds were destroyed by disease.

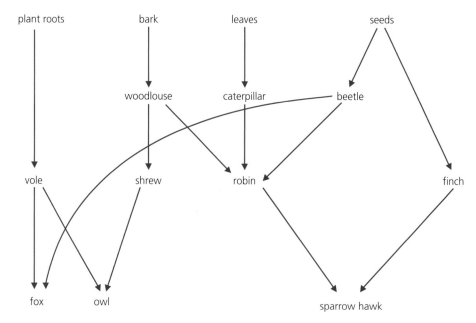

▲ **Figure 5.6** Example of a food web for a European woodland.

Population change

Ecologists carry out habitat surveys to find an estimate of the number of individuals of different species in a given area. For example, if 25 grass plants were found in a small square of a lawn, and it was estimated that the data from 100 squares would be needed to survey the whole lawn, an estimate of the population of the grass plants in the lawn could be made by multiplying the number of plants in one square by the number of squares needed to complete the survey.

9 If 25 grass plants are found in one square in the lawn and the lawn covers an area equal to 100 squares, what is the estimate number of grass plants in the lawn?

10 How reliable is the evidence provided by the estimate? Explain your answer.

11 One sweep of a sweep net in a grassy area collected seven spiders. Two more sweeps collected five and eight respectively. It was estimated that 80 sweeps of the net would be needed to collect all the spiders in the area.
   a How would you use this information to work out an estimate for the total number of spiders in the area?
   b What is your estimate of the spider population in the area?
   c How reliable is the evidence provided by the estimate? Explain your answer. What could you do if you if you decided it was not reliable?

Once data from a habitat survey has been collected, it is stored and a second survey is taken some time later to look for changes in the population. This may be repeated over many years to monitor population changes.

Sometimes populations of different species are recorded at the same time. Charles Elton, an English ecologist, was interested in how the populations of a **prey** animal and its predator might change over time. He did not survey a habitat, but instead used the records of the Hudson Bay Company of Canada. The company traded in animal pelts (their fur) provided by trappers. From this data, Elton extracted information about the number of snowshoe hares and lynx that were trapped over a 90-year period. From this data, he produced a graph, shown in Figure 5.7.

> **DID YOU KNOW?**
> Sudden changes shown on a line graph are called fluctuations. The highest point recorded on a line graph is called the peak.

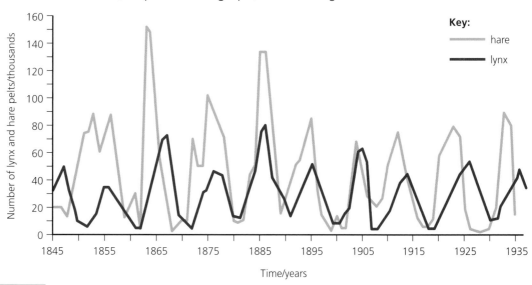

Key:
— hare
— lynx

▲ **Figure 5.7** The changes in the number of animal pelts supplied by trappers to the Hudson Bay company.

> **CHALLENGE YOURSELF**
> Find the peak number of hare and lynx pelts and state the year in which the peak was reached for each.

12 Does the graph show a trend or a pattern?

13 How does the population of snowshoe hares vary over 90 years?

14 How does the population of lynx vary over 90 years?

15 It has been suggested that the way the populations vary is due to predation. Do you agree? Explain your answer.

Predictions about changes in the population of a species can be made by finding out about the **reproductive rate** of the species (in mammals this is called the **birth rate**) and the **death rate**. In humans, the birth rate is the number of babies born per 1000 people (in a country) and the death rate is the number of people dying per 1000 people (in a country) in one year. When the birth rate is greater than the death rate, the population will increase in size. When the death rate is greater than the birth rate, the population will decrease in size.

Surveying methods are used to establish the size of a population of living things, either in a habitat or anywhere else throughout the world. A change of status in a population of living things in their habitat is due to a variety of causes, which may include predation, disease or environmental change. These can be natural changes, such as the effects of earthquakes or volcanic eruptions, but most of the changes today are due to the activities of humans, such as changes in agriculture and fishing, extraction of minerals and materials, waste products of industrial processes and forms of travel.

Endangered species

Data about world populations of living things is entered on the International Union for Conservation of Nature's (IUCN's) Red List of Threatened Species. This widely used list is divided into different categories and each species is entered in one of them. The table shows some of the categories with their symbol and a brief description.

▼ **Table 5.1** The path to extinction.

Status	Symbol	Description
Endangered	EN	High risk of extinction in part of the world or the whole world.
Critically endangered	CR	High risk of extinction in the wild.
Extinct in the wild	EW	The result of many surveys shows it no longer is present in the wild. It may survive in captivity or in cultivation.
Extinct	EX	No individuals have been recorded after many surveys.

16 What will happen to a population if the birth rate and the death rate are the same?

17 Go online and use the IUCN's Red List to see if you can find an example of a plant and an animal in your region of the world that is endangered or critically endangered. Compare your findings with those of others.

The IUCN's Red List of Threatened Species lists the species of plants and animals that are threatened with extinction today. In many cases, species are under threat as a result of environmental changes that have affected their habitats. Many of these environmental changes are caused by human activity. They include:
- cutting down rainforests to sell the trees (deforestation)
- using land that was a rainforest to grow crops, such as rubber or palm oil
- building houses on land where plants and animals live

CHALLENGE YOURSELF

Use the internet and other resources to find examples of environmental changes due to one or more of the following human activities: farming, fishing, extracting minerals, release of waste products, air travel and travel over land. Describe what the changes are and how they have affected the status of living things in the environment. Share your findings and make some suggestions for reducing or removing the environmental change.

- increased use of cars and other vehicles which produce carbon dioxide and other gases
- climate change as a result of burning fossil fuels
- climate change leading to changes in the temperature of the sea and sea levels, floods and droughts
- increased numbers of bush and forest fires
- pollution of the environment, especially by plastic which takes a long time to break down
- pollution of the environment by factories which produce harmful gases.

Many endangered animal species have been reduced to a small world population by hunting. Such animals have been killed more quickly than they can reproduce. If the death rate exceeds the birth rate, the species is set on a course for extinction.

Extinction means that a population of a species or a group of species no longer exists. All the individuals have died. This means that there are no individuals left to breed and keep the species in existence.

Endangered species can be helped by raising their birth rate and reducing their death rate. Zoos can help to increase the size of the world population of some endangered animal species. They increase the birth rate by ensuring that all the adult animals in their care are healthy enough to breed and by providing extra care in the rearing of the young. Zoos also reduce the death rate by protecting the animals from predation. In many countries, reserves have been set up in which endangered animals live naturally but are protected from hunting by humans. This reduces the death rate, which in turn increases the birth rate as more animals survive to reach maturity and breed.

LET'S TALK

▲ **Figure 5.8** The Javan rhinoceros.

The Javan rhinoceros was once numerous throughout Southeast Asia, but now only one small population of about 60 animals is known to exist for certain, in a protected area in Indonesia. The species is critically endangered and on the brink of extinction.

Large mammals, like the Javan rhinoceros, need large areas of natural habitat to support a big population. With the increasing human population, why is it difficult to conserve these large areas? Share your ideas in a group and explain your thinking.

Science extra: A surprising result may have saved a snail

In the past, purple dye for clothing was usually obtained by collecting and then boiling a marine snail, called the purple dye murex.

▲ **Figure 5.9** The purple dye murex snail.

▲ **Figure 5.10** Sir William Henry Perkin.

As time went on, the numbers of this snail became lower and lower, and it probably would have become extinct except for the surprising result of a chemical reaction.

William Perkin (1838–1907), an English scientist, was looking for a chemical cure for malaria. He knew that a compound called quinine, which could be extracted from certain trees in South America, was successful in treating the disease. However, this led to the trees being damaged. He thought he could make a chemical which had the properties of quinine. He experimented with a chemical called analine that is found in coal. It did not behave like quinine but produced a purple dye which could be used for colouring clothes. In a short time, this dye replaced the purple obtained from the snail and, thanks to this, populations of the purple dye murex snail still survive to this day.

Extinction events

An **extinction event** occurs when large numbers of species become extinct in the same period of time. Evidence for extinction events comes from studying **fossils** in the layers of rock laid down on the Earth over millions of years. Scientists have identified some possible causes for each of these events, but there is continuing debate about how each event occurred. They seem to be caused by significant environmental changes.

Table 5.2 provides a brief summary of five major extinction events, sometimes known as the 'Big Five'. The names of these events come from the geological time periods used to measure the Earth's history, starting with the Late Ordovician extinction event over 450 million years ago. The table also provides information about the sixth extinction event, which many scientists believe is occurring now. The timeline in Figure 5.11 shows some species that were living on Earth during the last 250 million years, and the extinction events that caused them to be destroyed.

▼ **Table 5.2** Summary of the major extinction events on Earth.

Extinction event	Time and duration (millons of years ago)	Possible causes and resulting extinctions
1 Late Ordovician	455–430	Ice age, volcanic action and weathering of rocks 85% of marine species lost
2 Late Devonian	376–360	Volcanic action, development of land plants, reduction of carbon dioxide 75% of all species lost
3 Permian–Triassic (also known as the 'Great Dying')	252	Volcanic action, **global warming** 96% of marine species, 70% of land vertebrates and large numbers of insect species lost
4 Triassic–Jurassic	201	Global warming, acidic oceans 80% of all species lost
5 Cretaceous–Tertiary	66	**Asteroid** 76% of all species (including the dinosaurs) lost
6 Anthropocene extinction	0.01	Large-scale human activity over the planet IUCN predicts more than 40 000 species are threatened with extinction, but in future this figure may increase.

18 Describe the environmental changes that caused, or are causing, each of the extinction events listed in Table 5.2.

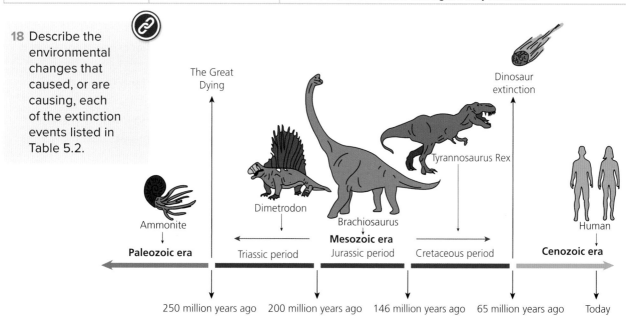

▲ **Figure 5.11** A timeline of species that lived on Earth.

Summary

✔ Organisms are adapted to the environment they live in and if the environment changes, it can affect the population of a species.
✔ Close observation of the environment can provide data to prevent a species from becoming extinct.
✔ Studies of population changes can identify species that are at risk of extinction.
✔ Extinction events are periods of time when large numbers of species become extinct.
✔ Jane Goodall observed chimpanzees in their natural environment over many years to discover how they have adapted and how they behave. Her work and later work with colleagues has increased our knowledge about the lives of chimpanzees.

CHALLENGE YOURSELF

Use the internet to find five animals that have become extinct in each of the Big Five extinction events.

LET'S TALK

How much have extinction events affected the animal life we have on the planet today? Explain your answer.

CHALLENGE YOURSELF

Ask people you know, or use the internet to find out, if there are any groups where you live who are trying to reduce environmental change. What are they trying to do? How successful are they? Which group would you support and what information helped you to decide?

End of chapter questions

1 What is an adaptation of a plant or an animal?
2 Name two ecological models that you have used to understand how food passes through an ecosystem.
3 If you put a bucket into a small pond, then brought it out and found four fish swimming in it, how could you estimate the population of fish in the pond?
4 If an organism in an ecosystem is endangered, what other stages will it go through before it becomes extinct?
5 How can close observation of a species, such as Jane Goodall's work with chimpanzees, help to prevent a species extinction?
 6 Look at all the survey techniques in this chapter. How would you use them to make sure that a species really has become extinct? How certain could you be of your conclusion?

 Now you have completed Chapter 5, you may like to try the Chapter 5 online knowledge test if you are using the Boost eBook.

The periodic table

In this chapter you will learn:
- about John Dalton and atomic weights (Science in context)
- about the first sorting of elements into groups (Science in context)
- how elements are arranged in the periodic table
- that the atomic number of an element depends on the atomic structure of the element
- how to use the periodic table to predict an element's structure
- about the physical and chemical properties of group 1 elements.

Do you remember?

- What is an **atom**?
- Describe the structure of the atom.
- Are all atoms the same? Explain your answer.
- What is an element?
- What is the periodic table?

▲ **Figure 6.1** Atomic structure describes what is going on inside an atom. Here we have atoms bonded together into a structure that we will talk about in the next chapter. The atoms of elements bond together in many ways to form compounds.

Sorting out elements

Science in context

John Dalton and atomic weights
John Dalton (1766–1844), the English chemist who constructed the atomic theory, also tried to put the elements in order. He measured the masses of the elements he collected when he broke up compounds. At the time, the term 'mass' was not used. The terms 'weight' and 'atomic weight' were used instead.

Dalton used the weight of hydrogen to compare with the weight of other elements. For example, when he separated hydrogen and oxygen from the compound water, he found that the weight of oxygen was seven times greater than the weight of hydrogen. He believed that all atoms combined with just one atom of another element, described as a ratio of 1 : 1. This meant that he thought the atomic weight of hydrogen was one, and the atomic weight of oxygen was seven. He used this idea to measure the atomic weights of other elements and set them out in a table.

Unfortunately, Dalton was not a very accurate experimenter, and other scientists discovered that the weight of oxygen produced when water splits up is eight times greater than the weight of hydrogen. Water is composed of two atoms of hydrogen and one atom of oxygen. This led them to work out that the atomic weight of oxygen is 16.

1 What was the inaccurate observation that Dalton made?

2 What inaccurate conclusion did Dalton make from his studies?

▲ **Figure 6.2** Chemical balance scales.

An example of early scientific equipment

In the early days of chemical studies, like those of Dalton, weights were measured using balance scales like the one in Figure 6.2.

There were many adjustments to be made before they could be used. For example, the levelling screws had to be used to make the balance perfectly level, and the plumb bob hanging down by the central support had to be used to make sure that the pointer was pointing directly vertical (down) on the midpoint of the scale. The instrument was used by putting the chemicals to be weighed in one pan and then weights were added to the other until the pointer returned to the midpoint position again.

3 How easy would it be to make errors using the balance scales shown in Figure 6.2? Explain your answer.

CHALLENGE YOURSELF

Can you make a set of balanced scales like the ones in Figure 6.2? Use your creativity to select items and materials to use. Make a labelled diagram of your balance and, if your teacher approves, make it. How can your previous knowledge of moments help you in the design of your balance?

 ## The first use of chemical symbols

John Dalton was the first scientist to use symbols in chemistry. He used circles to represent what he thought were elements. He joined circles of different elements together to represent the atoms in a compound.

4 'Azote' is another name for nitrogen and 'platina' is another name for platinum, but there are six names in Dalton's list that are actually compounds and not elements. Which are they?

5 Use secondary sources to find the symbols that we use today for the elements present in the list in Figure 6.3.

▲ **Figure 6.3** The symbols and atomic weights of substances that Dalton believed to be elements in 1805.

Science in context

▲ **Figure 6.4** Johann Wolfgang Döbereiner.

Classifying elements into groups

Johann Wolfgang Döbereiner (1780–1849), a German chemist, studied Dalton's work. In 1829, when more than 12 new elements had been discovered, he began sorting them out and found that he could divide them into groups of three according to their atomic weights and properties. These groups became known as Döbereiner's Triads. In one triad he placed lithium, sodium and potassium, and in another he placed chlorine, bromine and iodine, while in a third triad he placed calcium, strontium and barium.

Can you classify elements by their properties? Here are nine elements with information about their chemical and physical properties. Read the information and then classify the elements into three groups of three based on the information provided. Write down the letters of the elements in each of your three groups.

▼ **Table 6.1** The chemical and physical properties of nine elements.

Element	Properties
A	Soft, silvery, fizzes when added to water
B	Green gas, does not conduct electricity
C	Forms compounds that burn with a dark red (crimson) flame with a Bunsen burner
D	Red liquid, does not conduct electricity
E	Soft, silvery, sets on fire with water and may explode
F	Forms compounds that burn with a green/yellow flame with a Bunsen burner
G	Forms compounds that burn with a bright red flame with a Bunsen burner
H	Soft, silvery, fizzes when added to water and produces an orange flame
I	Black solid, does not conduct electricity

Looking for patterns in the properties of the elements

Another ten elements had been discovered by the time the English chemist John Newlands (1837–98) attempted to sort out the elements. Newlands set out the elements in order of atomic weights, starting with the lowest. As he examined the properties of the elements in his list, he noticed that some of the properties appeared **periodically** in elements that were eight places apart.

The names of these elements are written in order of their atomic weights, starting with the smallest and finishing with the largest: lithium, beryllium, boron, carbon, nitrogen, oxygen, fluorine, sodium, magnesium, aluminium, silicon, phosphorus, sulfur, chlorine, potassium and calcium.

Row 2	^{3} Li 6.94	^{4} Be 9.01	^{5} B 10.8	^{6} C 12.0	^{7} N 14.0	^{8} O 16.0	^{9} F 19.0	^{10} Ne 20.2	
Row 3	^{11} Na 23.0	^{12} Mg 24.3	^{13} Al 27.0	^{14} Si 28.1	^{15} P 30.1	^{16} S 32.1	^{17} Cl 35.5	^{18} Ar 39.9	
Row 4	^{19} K 39.1	^{20} Ca 40.1							

▲ **Figure 6.5** Rows 2, 3 and 4 of the periodic table.

6 In what way was Dalton's work useful as evidence to Döbereiner and Newlands?

7 Could Döbereiner and Newlands attempt a more detailed sorting out of the elements than Dalton because they had more data? Explain your answer.

▲ **Figure 6.6** Mendeleev as a young man working in a laboratory.

Look at rows 2, 3 and 4 from the periodic table (shown in Figure 6.5). Note the positions of lithium, sodium and potassium, which have similar properties. Now, count the number of elements from lithium to sodium and also from sodium to potassium. What do you notice?

Neon and argon were discovered after Newlands had completed his work. Neon was inserted between fluorine and sodium, and argon was inserted between chlorine and potassium.

Science extra: How the periodic table was made

Dmitri Mendeleev (1834–1907) was a Russian scientist who examined the work of Newlands and Dalton.

Following on from the work of John Newlands, Mendeleev began to sort out the elements. He arranged them in order of their atomic weights, but he also noticed something else – about the way the atoms joined together. Dalton thought that atoms only joined in a ratio of 1 : 1, but this was an error. It was discovered afterwards that the atoms of some elements were capable of joining with two (or three or more) atoms of other elements. When Mendeleev studied the elements in his list, he noticed that lithium would join with only one other atom, but that beryllium (next to it) would join with two atoms. The next element, boron, joined with three atoms and carbon, which followed, joined with four atoms. As he went down his list, he saw that the elements joined with three, two and one atoms respectively before there was another rise and fall, and so on.

◀ **Figure 6.7** Handwritten early version of Mendeleev's periodic table – you can see how he made changes as new data became available.

---

8 What evidence did Mendeleev consider when starting work on the periodic table?

9 How did Mendeleev's powers of observation help him in constructing the periodic table?

10 What creative thought did Mendeleev apply to the table when he found that some of the data did not fit to form a regular pattern?

11 How did Mendeleev test his table to see if it could be used to show the relationships of all elements?

Mendeleev rearranged the elements in his list into rows in a table, so that each row had a rise and fall in the number of atoms the elements would combine with. When he looked at the properties of the elements as they were arranged – one below the other in columns in the table, he noticed that they not only combined with the same number of atoms, but also had similar properties. As he looked along the rows in his table, he could see that the ability of the elements to join with atoms and the properties they possessed changed periodically.

However, there was a problem with the table – the data did not all fit in to create a regular pattern in some places. Mendeleev knew that many elements had been discovered in the nineteenth century and there was a chance that there were more still to be discovered. He therefore left gaps in his table that could be filled as the elements were discovered. This meant that he could use the table in another way – to look at the elements around the gaps and predict the properties of the unknown elements. Other scientists did not approve of the idea of leaving gaps. They thought that it was cheating and that it was done to support his theory, which they found unbelievable. However, they had to change their minds when new elements *were* discovered that had the properties Mendeleev predicted, and these elements were added to the table.

The periodic table and atomic number

Earlier we saw that chemists in the nineteenth century used the term 'atomic weight' in sorting out the elements. Mendeleev's **periodic table** was a model of how the elements could be arranged to help him predict the properties of elements that were yet to be discovered. It used atomic weights as Dalton had done in his atomic theory. Today, the periodic table uses **atomic numbers** instead of atomic weights, but it is still a model that we can use to see how elements are arranged, and to predict their properties.

> **DID YOU KNOW?**
> The periodic table gets its name from the horizontal rows of elements in the table. These rows are called 'periods'.

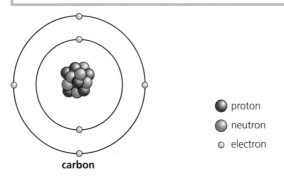

proton
neutron
electron

carbon

▲ **Figure 6.8** A carbon atom.

The atoms of each element contain a nucleus with a certain number of **protons** which are unique to that element (see Figure 6.8). The number of protons in an atom is called the **atomic number** and this is now used to arrange the elements in the periodic table. The atomic number of each element can be seen in the top left corner of the box for that element.

1 H hydrogen																	2 He helium
3 Li lithium	4 Be beryllium											5 B boron	6 C carbon	7 N nitrogen	8 O oxygen	9 F fluorine	10 Ne neon
11 Na sodium	12 Mg magnesium											13 Al aluminium	14 Si silicon	15 P phosphorus	16 S sulfur	17 Cl chlorine	18 Ar argon
19 K potassium	20 Ca calcium	21 Sc scandium	22 Ti titanium	23 V vanadium	24 Cr chromium	25 Mn manganese	26 Fe iron	27 Co cobalt	28 Ni nickel	29 Cu copper	30 Zn zinc	31 Ga gallium	32 Ge germanium	33 As arsenic	34 Se selenium	35 Br bromine	36 Kr krypton
37 Rb rubidium	38 Sr strontium	39 Y yttrium	40 Zr zirconium	41 Nb niobium	42 Mo molybdenum	43 Tc technetium	44 Ru ruthenium	45 Rh rhodium	46 Pd palladium	47 Ag silver	48 Cd cadmium	49 In indium	50 Sn tin	51 Sb antimony	52 Te tellurium	53 I iodine	54 Xe xenon
55 Cs caesium	56 Ba barium	57–71	72 Hf hafnium	73 Ta tantalum	74 W tungsten	75 Re rhenium	76 Os osmium	77 Ir iridium	78 Pt platinum	79 Au gold	80 Hg mercury	81 Tl thallium	82 Pb lead	83 Bi bismuth	84 Po polonium	85 At astatine	86 Rn radon
87 Fr francium	88 Ra radium	89–103	104 Rf rutherfordium	105 Db dubnium	106 Sg seaborgium	107 Bh seaborgium	108 Hs bohrium	109 Mt hassium	110 Ds darmstadtium	111 Rg roentgenium	112 Cn copernicium	113 Nh nihonium	114 Fl flerovium	115 Mc moscovium	116 Lv livermorium	117 Ts tennessine	118 Og oganesson

57 La lanthanum	58 Ce cerium	59 Pr praseodymium	60 Nd neodymium	61 Pm promethium	62 Sm samarium	63 Eu europium	64 Gd gadolinium	65 Tb terbium	66 Dy dysprosium	67 Ho holmium	68 Er erbium	69 Tm thulium	70 Yb ytterbium	71 Lu lutetium
89 Ac actinium	90 Th thorium	91 Pa protactiniu	92 U uranium	93 Np neptunium	94 Pu plutonium	95 Am americium	96 Cm curium	97 Bk berkelium	98 Cf californium	99 Es einsteinium	100 Fm fermium	101 Md mendelevium	102 No nobelium	103 Lr lawrencium

▲ **Figure 6.9** The whole of the modern periodic table.

12 How are the elements arranged by their atomic number – horizontally or vertically?

Look at the periodic table to help you to answer these questions.

13 What is the atomic number of
- a calcium
- b phosphorous
- c chlorine
- d potassium?

14 How many protons are there in each of these elements?
- a magnesium
- b oxygen
- c carbon

DID YOU KNOW?

Hydrogen is placed on its own in the periodic table because its properties do not match well with the properties of the other elements. It is also unusual because it does not have a **neutron** in its nucleus – just one proton surrounded by one **electron**.

Group 1 in the periodic table

The periodic table can be used to predict an element's structure and its physical and chemical properties.

Many of the columns of elements in the periodic table are called **groups**. The elements in a group share similar properties. A trend can be seen in the properties as you go down the group.

Group 1 in the periodic table contains the alkali metals. The physical properties of the alkali metals are shown in Table 6.2.

▼ **Table 6.2** The physical properties of the alkali metals.

Element	Density/g per cm³	Melting point/°C	Boiling point/°C
lithium	0.53	180.6	1344
sodium	0.97	97.9	884
potassium	0.86	63.5	760

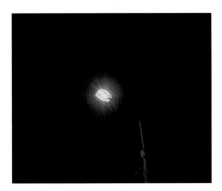
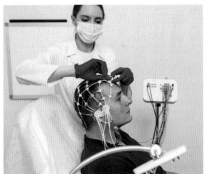
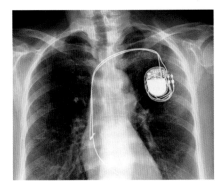

▲ **Figure 6.10** Sodium in a street lamp, potassium nodes on a brain scanner and a lithium battery in a pacemaker.

15 Study Table 6.2 alongside the periodic table on the previous page, and identify
  a any trend
  b any patterns
  c any anomalous results.

16 Is there a trend in which the alkali metals react with water? Explain your answer.

Chemical properties of the alkali metals

When samples of the alkali metals are added to water in turn, the following reactions take place:
● Lithium fizzes as it floats on the water.
● Sodium fizzes more strongly.
● Potassium bursts into flame.

In all three reactions, the metal hydroxide is produced in the liquid and hydrogen gas escapes into the air. In the reaction with potassium and water, the heat produced causes the hydrogen to burn in the air and produce a flame. When the alkali metals burn in air, they produce solid metal oxides.

17 From the way alkali metals behave with water, how do you think the strengths of the reactions with oxygen will also show a trend? Explain your answer. (**Hint:** think about the **reactivity series**.)

CHALLENGE YOURSELF

Create and design a simple interactive game about the elements in group 1. Set it out and let your friends try it. How does it help you learn about the elements?

Summary

✔ The arrangement of elements in the periodic table is based on their atomic structure.
✔ The number of protons in an atom is the atomic number.
✔ Elements in the same group (for example, group 1) have similar chemical and physical properties.
✔ Over time, different individuals used their scientific understanding of different elements to see if there was a way to organise them based on their properties, and this developed into the periodic table we use today.

End of chapter questions

1 The following scientists worked to establish an order in the elements. Arrange their names in the order in which they would appear on a timeline.
   – Mendeleev
   – Döbereiner
   – Dalton
   – Newlands

2 What piece of equipment did Dalton use in his work on atomic weights?

3 a What does the atomic number of an element tell you about the nucleus of its atom?

   b Why can the atomic number be used to identify an element?

4 Sodium is a softer metal than lithium. Describe how you think the softness of potassium compares with that of sodium.

5 Which metal has the smallest temperature range when it is in liquid form? (Look back at the Table 6.2 on page 68 to help you.)

 6 How does the periodic table help in the study of chemistry?

 Now you have completed Chapter 6, you may like to try the Chapter 6 online knowledge test if you are using the Boost eBook.

Bonds and structures

In this chapter you will learn:
- about the electrons in an atom
- how covalent bonds form when the atoms in a reaction share the electrons in their outer orbit
- how covalent bonds form with oxygen (Science extra)
- that when atoms join together, they form molecules
- about the use of molecular modelling in chemical research (Science in context)
- how the bucky-ball molecule was discovered (Science extra)
- how some atoms can lose an electron from their outer orbit to become positively charged ions, and other atoms can gain an electron in their outer orbit to become negatively charged ions
- how to describe ionic bonds as the attraction between positive ions and negative ions
- how to explain the difference between simple structures and giant structures.

Do you remember?

- What is the difference between an element and a compound?
- Ernest Rutherford made a model of the atom. Describe it.
- What do you find around the nucleus in an atom?
- Which parts of atoms are:
  - a positively charged
  - b negatively charged
  - c do not have a charge?

The electrons in an atom

There are a number of models of the arrangement of electrons in an atom that scientists use. The first one to consider is called the **electron cloud**. It describes where the electrons are inside the atom.

A second description of the atom, based on further research, features the movement of electrons in **orbits** around the nucleus. Some electrons are in orbit closer to the nucleus than others, as the orbital model for the beryllium atom in Figure 7.2 on the next page shows.

▲ **Figure 7.1** The electron cloud.

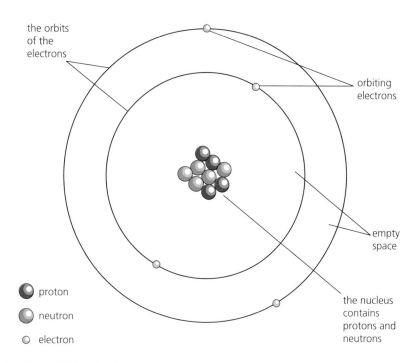

the orbits of the electrons

orbiting electrons

empty space

the nucleus contains protons and neutrons

proton

neutron

electron

▲ **Figure 7.2** The basic structure of an atom. This example is a beryllium atom.

Be

▲ **Figure 7.3** The atomic structure of beryllium.

He

▲ **Figure 7.4** The atomic structure of helium.

Ne

▲ **Figure 7.5** The atomic structure of neon.

A third description of the atom presents the detail of the nucleus and the orbits of the electrons more simply, as Figure 7.3 shows.

In Figure 7.3, the number of protons (p) and neutrons (n) is shown in the centre of the diagram, representing the nucleus, and the electrons (e) are shown in their orbits, around the centre. The protons, neutrons and electrons in an atom are called sub atomic particles.

Each orbit can only hold a certain number of electrons before it is full. For example, a helium atom, which has one electron shell, can only hold two electrons and then it is full.

Other atoms have an orbit of two electrons surrounded by more orbits. In the beryllium atom, for example, there is just one more orbit containing two electrons. This does not mean that this orbit is full, however, because some elements need eight electrons in their orbits before they are full, and some elements need even more.

The orbit of electrons which is most important for chemical reactions is the outer orbit. When this orbit is full, the atom becomes unreactive and is described as stable. Elements that naturally have full orbits are the inert or noble gases, such as helium, with its full orbit of two electrons, and neon (see Figure 7.5) with a full orbit of eight.

Covalent bonding

When two or more atoms join together, they form molecules. Some molecules are made up of two atoms of the same element; for example, molecules of hydrogen and oxygen. Other molecules contain different elements; for example, a molecule of water contains hydrogen and oxygen atoms. A protein molecule contains hydrogen, nitrogen, carbon and oxygen atoms.

There are different types of bond that hold the atoms together in a molecule. The bond between electrons in the outer orbit is called the **covalent bond**. This bond forms when two atoms take part in a chemical reaction and share one or more electrons in their outer orbits. When they do this, they fill the outer orbit with electrons and become stable, like the atom of an inert gas.

Hydrogen

A hydrogen atom has only one electron in its one shell, but it needs two for stability. It achieves this by forming a covalent bond with another hydrogen atom, as Figure 7.6 shows. To make the sharing of electrons clearer, a different symbol for an electron is used in each of the two atoms. When the two electrons are shared, the symbols for the two electrons are placed in the diagram in a space where the two orbits overlap.

two hydrogen atoms

H   H

a hydrogen molecule, H_2

H   H

a shared pair of electrons

▲ **Figure 7.6** The covalent bond of hydrogen.

Covalent bonds also form between atoms of different elements to create compounds.

Water

A water molecule is composed of two atoms of hydrogen and one atom of oxygen. They form covalent bonds as shown in Figure 7.7.

1 Describe how the oxygen atom and the two hydrogen atoms become stable through the covalent bonds.

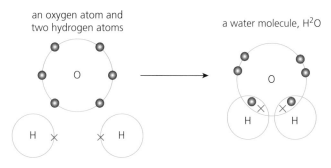

an oxygen atom and two hydrogen atoms

a water molecule, H^2O

▲ **Figure 7.7** The covalent bond of water.

Methane

A molecule of methane (the gas used in a Bunsen burner, if that is a heat source found in your school laboratory) is made from one carbon atom and four hydrogen atoms. They form covalent bonds as Figure 7.8 shows.

a carbon atom and four hydrogen atoms

a molecule of methane, CH_4

2 Describe how the carbon atom and the four hydrogen atoms become stable through the covalent bonds.

▲ **Figure 7.8** The covalent bond of methane.

CHALLENGE YOURSELF

Make models of:
a   a hydrogen atom and molecule
b   a water molecule
c   a methane molecule.

Decide what to use for the outer orbit and the electrons and how you will show the bond. Then, if your teacher approves, collect your materials and items and make your models.

Science extra: Oxygen

O

▲ **Figure 7.9** An oxygen atom.

When studying bonds, only the outer orbit is represented in a diagram of an atom. When studying how the oxygen atoms form a covalent bond, only the outer orbit is represented, as shown in Figure 7.9.

When an oxygen atom forms a covalent bond with another oxygen atom, it shares two electrons. This means that each atom now has eight electrons in its outer orbit instead of six and each atom becomes more stable.

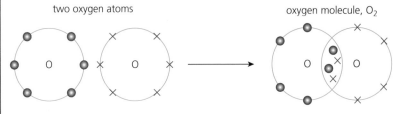

two oxygen atoms

oxygen molecule, O_2

▲ **Figure 7.10** The double covalent bond of oxygen.

Molecular models

John Dalton used circles to represent atoms of elements. His idea was later developed into three-dimensional models by the German chemist August Wilhelm von Hofmann (1818–92). He used croquet balls and brass tubes to make his models. Hofmann also invented a system of colouring the balls to represent atoms of different elements.

▲ **Figure 7.11** A modern 3-D molecular model of methane, created in the style of Hofmann's original model.

This system was further developed by other scientists and is now called CPK colouring. It is named after the scientists who developed the molecular models: Robert Corey, Linus Pauling and Walter Koltun.

Figure 7.12 on the next page shows part of the CPK colour-code system used today. You can use the same colour codes when making your own molecular models.

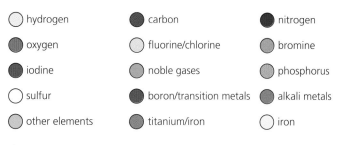

○ hydrogen	● carbon	● nitrogen
● oxygen	○ fluorine/chlorine	● bromine
● iodine	● noble gases	● phosphorus
○ sulfur	● boron/transition metals	● alkali metals
● other elements	● titanium/iron	○ iron

▲ **Figure 7.12** CPK colouring.

Science in context

The use of molecular modelling in chemical research

A molecular model allows scientists to see the position and relationships between the atoms that make up a molecule. When scientists examine the structure of the molecular model and how the molecules of a substance behave in chemical reactions, they can link the structure to the properties.

▲ **Figure 7.13** Scientists using 3-D glasses to view the molecule shown on the computer screen. The glasses receive information from an infrared beam that is directed to them from the small box on top of the computer.

When they do this, they can make predictions about what might happen to the chemical properties if the arrangement of the atoms in the molecule were changed, or if similar atoms were added or subtracted to make new molecules. For example, a particular molecular structure may give the substance it forms a particular chemical property which is useful in attacking a disease-causing organism. Using previous knowledge and understanding about elements and molecular structures, a new molecule can be made by adding one or more atoms of an element. This helps the substance to more effectively attack the disease-causing organism, bringing the disease under further control.

You have tried some activities of modelling molecules in three dimensions using objects you can hold and handle, but scientists also use computers to help them model complicated molecules and view them from all angles to help them in their studies.

CHALLENGE YOURSELF

Create a presentation that describes molecular models and their uses to a group of people who have become familiar with atomic structure. How will you illustrate your presentation?

3 How might a molecular model help in chemical research?

4 What is the advantage to using computer modelling over making models with objects you can hold?

Science extra: An unexpected discovery about carbon molecules

At the beginning of the 1980s, the chemist Harry Kroto (1939–2016) was working with his students on molecules of carbon. The molecules formed long chains and the scientist observed them bending and turning. Other research in space had shown that similar molecules of carbon were found in a gas cloud around a star called Taurus. Harry believed that the star was making the molecules and he set out with other colleagues to try and make the molecules in a similar way on Earth. A **laser** beam was fired at a piece of graphite and the molecules were produced. However, something else was produced that was unexpected – a molecule of sixty carbon atoms with an unknown structure. The scientists worked on constructing a model. At first, they used toothpicks and jellybeans, then, as they continued to brainstorm and produce more ideas, they used paper cut-outs. Eventually they discovered that the molecule was shaped like a ball. The molecule is called a bucky-ball and was named after the American architect Buckminster Fuller, who designed domed structures like the one shown here.

▲ **Figure 7.14** The Montreal Biosphere.

Ions and ionic bonding

A simple ionic bond forms when one atom gives one electron away to another atom to make itself stable. The atom which receives the electron also becomes stable. The movement of an electron between a sodium and a chlorine atom is shown in Figure 7.15 on the next page.

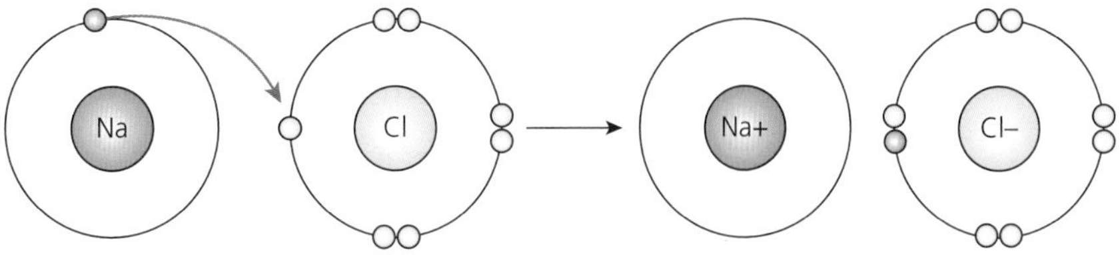

▲ **Figure 7.15** The movement of an electron between sodium and chlorine.

When the sodium atom loses an electron, the charge on the atom changes because there are still eleven positively charged protons in the nucleus, but now only ten electrons surrounding it. This makes the atom with its lost electron positively charged.

When the chlorine atom gains an electron, the charge on the atom changes because there is now an extra electron in the group surrounding the protons in the nucleus. As the electron carries a negative electrical charge, it makes the atom negatively charged.

An atom which becomes charged is called an **ion**. An ion with a positive charge like the sodium ion is called a **cation**. An ion with a negative charge like the chlorine ion is called an **anion**.

$[Na]^+$ $[:\ddot{C}\ddot{l}:]^-$

▲ **Figure 7.16** The ionic bond that exists between sodium and chlorine.

The ions are attracted by their equal, and opposite, electrical charges. They can be represented in a diagram with brackets [ ] around each ion, together with the charge that it possesses as + or −. Thus, the ionic bond that exists between sodium and chlorine can be represented in a diagram, as shown in Figure 7.16.

In a covalent bond, more than one electron is shared. In a similar way, more than one electron may pass from one atom to the next in ionic bonding.

Structures

The atoms of elements and compounds form structures. These structures can be simple and made from only a small number of atoms, or they can be giant structures and made from a huge number of atoms.

Simple structures

The molecules of hydrogen, water and methane are **simple structures**. They are made from a small number of atoms held together by covalent bonds. The molecules do have small forces which can hold them together, with more of the same molecules, to make liquids or solids if they are cold enough. These forces are so weak that when energy is given to a collection of these simple molecules, it breaks these forces between them and the molecules move apart. Simple structures have low melting and boiling points. For example, hydrogen melts at −259.2 °C and boils at −252.9 °C and methane melts at −182.5 °C and boils at −162 °C. As all the electrons are tightly bonded between the atoms, simple molecular structures do not conduct electricity.

Giant covalent structures

The huge number of atoms that form **giant covalent structures** create a repeating pattern of linkages called a **lattice**. Diamond and graphite are examples of giant structures that are formed in this way. We use graphite in pencils and diamonds in jewellery and some tools. Diamond is formed by the carbon atoms sharing four electrons to make the bonds, while graphite is formed using only three of their electrons to make covalent bonds.

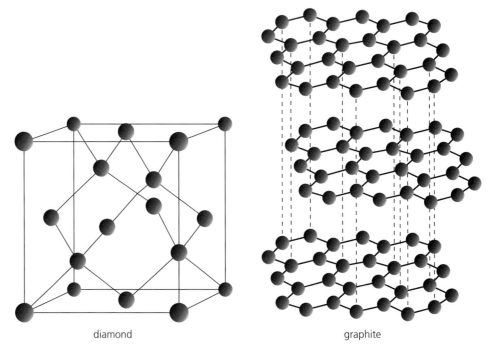

diamond   graphite

▲ **Figure 7.17** The giant covalent structures of **a** diamond and **b** graphite are formed by a lattice of carbon atoms.

DID YOU KNOW?

▲ **Figure 7.18** A graphite pencil (the car has been drawn by a graphite pencil, using thick lines) completing a circuit to power a lamp.

As the carbon atoms in graphite only use three of their electrons to form bonds, this means that each atom has one electron that is unused and is free to move around the structure. These take part in conducting electricity that flows through the material. It is these free electrons that make graphite a good electrical **conductor**. It is so good, a pencil can be used to complete an electrical circuit to power a small fan or to light an LED bulb!

▲ **Figure 7.19** Quartz rock crystal.

Silica is another example of a giant covalent structure which is formed from the atoms of two elements, silicon and oxygen. It forms a mineral called quartz.

The bonds that hold giant covalent structures together are very strong. When all the electrons are used in bond formation, the materials (like diamond and silica) are very hard. As there are no free electrons, they are also non-conductors of electricity. The very strong covalent bonds that hold the atoms together need a great deal of energy to break them and they all have high melting points. For example, graphite breaks down at 3600 °C but does not melt – it sublimes, which is a process where a solid changes directly into a gas.

> 5   What does the information about giant covalent structures tell you about the atoms which make them? Do you think they are metals or non-metals?

Giant ionic structures

Atoms which release or receive electrons become electrically charged ions. The force generated by the electrical charges holds them together. They do not form molecules but they do join together, creating different structures.

 In Figure 7.20, you can see that the ions of sodium and chlorine pack tightly together. If you were to count them all, you would find that there was an equal number of sodium and chlorine ions. This arrangement is expressed as the formula unit NaCl.

The structure for magnesium oxide is shown in Figure 7.21. If you were to go through a lattice of magnesium oxide, counting the atoms of magnesium and oxygen, you would find that for every ion of oxygen, there is an ion of magnesium. This arrangement is expressed as the formula unit MgO.

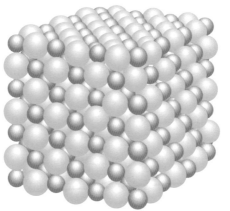

▲ **Figure 7.20** The structure of sodium chloride (salt).

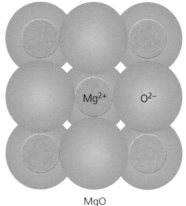

MgO

▲ **Figure 7.21** The structure of magnesium oxide.

The ionic bonds that hold the ions together in a lattice structure are so strong that they also need a great deal of energy to break them apart. As such, substances with these structures have very high melting and boiling points, because all the electrons are held within the bonds and there are none left to move freely. This also makes substances made from these structures non-conductors of electricity. However, if the lattice is placed in water and dissolves, then the ions can split apart and move freely. Then, as they are electrically charged, the watery solution can conduct electricity.

How can we use electricity to compare substances with covalent and ionic bonds?

Distilled water is pure water. It only contains water molecules which are formed by covalent bonds. Sodium chloride (common salt) crystals are structures formed by ionic bonds, but dissolve in water to form anions and cations (negative and positive ions).

Hypothesis
Construct a testable hypothesis from the information above and in the chapter to show how electricity can be used to detect substances which are electrical **insulators** and conductors.

Prediction
Construct a prediction base on your hypothesis.

Planning, investigating and recording data
1   Set out a plan based on your knowledge of testing for electrical insulators and conductors.
2   Make a list of equipment that you will need to carry out your plan.
3   Set out how you will record your observations and think about how many measurements you need to make.
4   Show your plan, equipment list and method of recording to your teacher and, if they are approved, try your investigation.

Examining the results
Examine the record of your observations and make comparisons.

Conclusion
Compare your results with your hypothesis and predictions and draw a conclusion.

Are there any limitations to your conclusion?

Are there further investigations which would help with your conclusion?

Summary

✔ When two or more atoms join together, they form molecules.
✔ Bonds hold atoms together in molecules.
✔ Covalent bonds occur when the atoms share the electrons in their outer orbits.
✔ Some atoms become negatively charged ions if they gain an electron in their outer orbit, or become positively charged ions if they lose an electron from their outer orbit.
✔ The attraction between negatively charged ions and positively charged ions results in ionic bonding.
✔ Simple structures have a small number of atoms held together by covalent bonds.
✔ Giant structures have large numbers of atoms held together by either covalent or ionic bonds.

End of chapter questions

1 How are electrons arranged around a nucleus?
2 Describe the arrangement of **subatomic particles** in a helium atom.
3 How does a covalent bond form?
4 How does an ionic bond form?
5 Why do atoms of different elements form bonds?
6 How are a giant covalent structure and a giant ionic structure
   a similar
   b different?

Now you have completed Chapter 7, you may like to try the Chapter 7 online knowledge test if you are using the Boost eBook.

In this chapter you will learn:
- what is meant by 'density'
- how to calculate the densities of solids, liquids and gases
- how to compare the densities of different materials
- to use density to explain floating and sinking
- about balloons and scientific research (Science in context).

Do you remember?

- What do scientists mean by the term 'mass'?
- What is the force of attraction that pulls the mass of your body towards Earth?
- What is the volume of an object?
- What piece of scientific equipment would you use to measure the volume of a liquid?
- What properties of objects affect whether they float or sink?

LET'S TALK

Which is heavier – the wood in the trunk of a tree or the metal in a coin? How can we test this accurately? Discuss your ideas in a group, then read the suggestion below.

Your first thought might be to say wood in the trunk of the tree is heavier, but consider that wood can float on water, while a coin would quickly sink. For there to be a fair comparison, the masses of equal volumes of the wood and the metal the coin is made from must be found. If we find the mass of a piece of wood the same size as a coin, we can see that the wood is in fact lighter than the coin.

▲ **Figure 8.1** Timber can be transported by water because wood is less dense than water, so it floats at the surface.

Defining and comparing density

The **density** of a substance is a measure of the amount of matter that is present in a certain volume of that substance. The following equation shows how the density of a substance can be calculated:

$$\text{density} = \frac{\text{mass}}{\text{volume}}$$

The basic SI unit of density is found by dividing the unit of mass by the unit of volume, so it is kg/m^3. This is pronounced 'kilograms per metre cubed'.

In the school laboratory, when small amounts of materials are used, the density of a substance is often calculated using masses measured in grams and volumes measured in centimetres cubed, giving a density value in g/cm^3. The density value in units of g/cm^3 can be converted to a value in kg/m^3 by multiplying it by 1000. For example, ice was found to have a density of $0.920\,g/cm^3$. This can also be expressed as $0.920 \times 1000 = 920\,kg/m^3$.

Table 8.1 shows the density of some of the most common solid materials.

▼ **Table 8.1** The density of common solids.

Material	Density/kg per m³
ice	920
cork	250
wood	650
steel	7900
aluminium	2700
copper	8940
lead	11350
gold	19320
polythene	920
perspex®	1200
expanded polystyrene	15

Density in solids

Measuring the density of a rectangular solid block

- The mass of a block is found by placing the block on a balance (check the balance reads zero first) and reading the scale. The mass is recorded in grams, g.
- The volume is found by multiplying the length, width and height of the block together and recording the value in cm^3.
- The density of the material in the block is then found by dividing the mass by the volume and expressing the quantity in the unit g/cm^3.

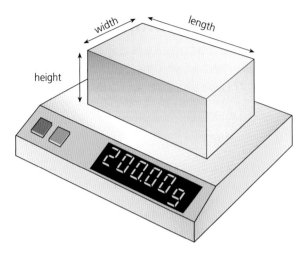

▲ **Figure 8.2** Finding the mass of a block using a top-pan balance.

1   A block of material is 8 cm long, 2 cm wide and 3 cm high, and has a mass of 46 g. What is its density?

2   a   Convert the value you found for the density in Question 1 to kg/m^3.
    b   Compare the density of the material in the block with those in Table 8.1 on page 83. Which materials in the table have densities closest to that of the block?
    c   How could you convert the value of a density given in kg/m^3 to g/cm^3?

How accurate are your estimates of density?

You will need:

blocks of wood, wax, plastic, steel, aluminium, and any other material that can be made into a block (the blocks may be of different sizes), access to a top-pan balance, a ruler.

Hypothesis
Construct a testable hypothesis relating to a way of estimating the density of a substance by feeling its weight.

Prediction
Construct a prediction based on your hypothesis.

Planning, investigating and recording data
Construct a plan that involves estimating the density of the blocks and recording the estimates. Then, using the information about measuring the density of the rectangular block, continue with the plan to calculate their actual densities and record them.

Examining the results

Compare the data provided by estimates with the data of measured densities.

Conclusion

Compare your evaluations with your hypothesis and prediction and draw a conclusion.

Is your conclusion limited in some way? Explain your answer.

What improvements could be made? Explain the changes that you suggest.

 ## Measuring the density of an irregularly shaped solid

The density of an irregularly shaped solid, such as a pebble, can be found in the following way:

- The mass of the pebble is found by placing it on a top-pan balance, just as for a solid of a regular shape.
- The volume is found by pouring water into a measuring cylinder until it is about half-full. The volume of the water is read on the scale and then the pebble is carefully lowered into the water on a thin string. When the pebble is completely immersed in the water, the volume of the water is read again on the scale. The volume of the pebble is found by subtracting the first reading from the second.
- The density of the pebble is then found by dividing the mass of the pebble by its volume.

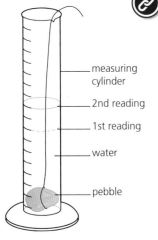

▲ **Figure 8.3** Measuring the volume of a pebble.

3 The mass of a pebble was 88.4 g. The original volume of water in the measuring cylinder was 50 cm³ and the combined volume of water and pebble was 84 cm³. What is the density of the rock that made the pebble?

Can you find the density of a pebble?

You will need:

a pebble that fits in a measuring cylinder, a measuring cylinder, string, a weighing machine such as a top-pan balance.

Planning, investigating and recording data

From your knowledge of volume and mass, work out a step-by-step way to answer the enquiry question.

If your teacher approves, try it.

How will you record your observations and measurements?

CHALLENGE YOURSELF

Can your technique be used to compare the densities of pebbles formed from different types of rock? Discuss your ideas with your teacher and, if approved, try them.

Density in liquids

Measuring the density of a liquid

The density of a liquid is found in the following way:

- A measuring cylinder is put on a balance and its mass is found (*A*).
- The liquid is poured into the measuring cylinder and its volume is measured (*V*).
- The mass of the measuring cylinder and the liquid it contains is found (*B*).
- The mass of the liquid is found by subtracting *A* from *B*.
- The density of the liquid is then calculated by dividing the mass of the liquid by its volume:

 $$\text{density} = \frac{B - A}{V}$$

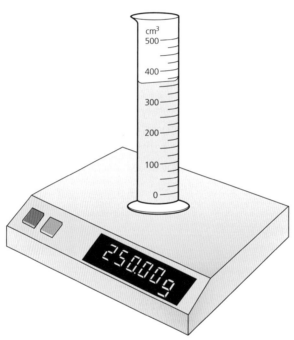

▲ **Figure 8.4** Finding the mass of a measuring cylinder and the liquid it contains.

Table 8.2 shows the density of some common liquids.

▼ **Table 8.2** The density of common liquids.

Liquid	Density/kg per m³
mercury	13550
water at 4°C	1000
corn oil	900
turpentine	860
paraffin oil	800
methylated spirits	790

Do different liquids have similar densities?

You will need:

a selection of liquids such as vegetable oil, sunflower oil, distilled water, a salt solution, honey, syrup, treacle, tomato ketchup.

Hypothesis

All liquids flow, therefore they have similar densities.

Is the hypothesis testable? Explain your answer.

Prediction

Make a prediction based on this hypothesis.

Planning, investigating and recording data

Set out a plan, using the five steps that are listed on page 86 to help you, and decide on a method of recording data.

Examining the results

Compare the densities of the liquids.

Conclusion

Compare your evaluation with your hypothesis and prediction and draw a conclusion.

Is your conclusion limited in some way? Explain your answer.

What improvements could be made? Explain the changes that you suggest.

▲ **Figure 8.5** Liquids of different densities form layers when they are mixed.

An unexpected result when ice melts

In the introduction to this book, you were invited to construct a hypothesis and prediction and make an investigation into ice melting in a glass filled to the brim with water.

Most people would probably say something like this: 'As ice is made of water and some of it is sticking out above the surface when the ice melts, this extra water will make the water in the glass spill over the brim.'

You should have found that when the ice is left to melt, this does not happen. The volume of water stays the same. This result is unexpected and needs further explanation.

 When a model of water molecules in ice is examined, it is seen that there are spaces between the molecules. These add to the volume of the ice. When the ice melts, these spaces no longer exist and the volume of the water that makes the ice is reduced so much that the water slips into the rest of the liquid without pushing the water over the brim.

Floating and sinking

When a piece of wood is placed in water, the wood floats. This is due to the difference in the densities of the wood and the water. When two substances are put together, such as a solid and a liquid or a liquid and a liquid, the less dense substance floats above the denser substance. When full-fat milk is poured into a container such as a bottle, the cream, which contains fat and is less dense than the more watery milk, rises to the top.

4 When paraffin oil and water are poured into a container, they separate and the paraffin oil forms a layer on top of the water.

When water and mercury are mixed, the water forms a layer on top of the mercury.
  a What can you conclude from these two observations?
  b What do you predict would happen if water and corn oil were mixed together? Use Table 8.2 on page 86 to help you.

5 What do you think would happen if the following solids were placed in water? Explain your answers.
  a expanded polystyrene
  b polythene
  c perspex®

6 What do you think would happen if the following solids were placed in mercury? Explain your answers.
  a steel
  b gold
  c lead

7 Why do you think the temperature of water is shown when the value of its density is given?

8 Most people can just about float in water (Figure 8.6). What does this tell you about the density of the human body?

9 When salt is dissolved in water, the solution that is produced has a greater density than pure water. An object that floats on pure water is shown in Figure 8.7. When it is placed in the salt solution, do you predict that it will rise higher than it did in pure water, or sink lower?

▲ **Figure 8.6** Humans can float in water.

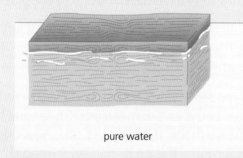

pure water

▲ **Figure 8.7** Pure water.

How can you compare the densities of liquids using a drinking straw, a lump of modelling clay and a marker pen?

You will need:

a drinking straw, a lump of modelling clay, a marker pen, a beaker of water, a beaker of saltwater, a beaker of vegetable oil.

Planning, investigating and recording data

Devise a piece of simple technology to answer the question and use it to test the liquids. Record your results.

Examining the results

Examine your data and compare it with some of the data you collected in the previous enquiry on the density of liquids.

Conclusion

Compare your evaluations with the question and conclude whether it has been answered or not.

Is your conclusion limited in some way? Explain your answer.

What improvements could be made? Explain the changes that you suggest.

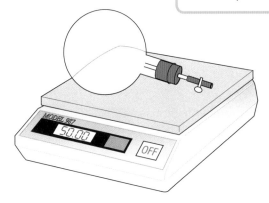

▲ **Figure 8.8** Finding the mass of a flask of air using a top-pan balance.

10 How is the process of finding the mass of a gas different from that of finding the mass of a liquid? Why is the difference necessary?

Density in gases

Air is a mixture of gases. Its density can be found in the following way:

- The mass of a round-bottomed flask with its stopper, pipe and closed clip is found by placing it on a sensitive top-pan balance. The flask is then attached to a vacuum pump and the air is removed from the flask and the clip is closed.
- The mass of the flask (with the air removed), stopper, pipe and closed clip is found by placing it back on the balance. The mass of the air in the flask is found by subtracting the second reading from the first.
- The volume of the air removed is found by opening the clip underwater so that water enters to replace the vacuum. The water is then poured into a measuring cylinder to find the volume.

The density of a gas changes as its temperature and **pressure** change. The densities of gases are compared by measuring them at the same

▼ Table 8.3 The density of common gases.

Gas	Density/ kg per m³
hydrogen	0.089
air	1.29
oxygen	1.43
carbon dioxide	1.98

temperature and pressure. This is called the **standard temperature and pressure** (STP). The standard temperature is 0 °C. The standard pressure of a gas is the pressure that will support 760 mm of mercury in a vertical tube. When two gases meet, the less dense gas will rise above the denser gas.

Table 8.3 shows the densities of some gases.

Use the table to help you answer these questions.

11 The atmosphere on Mars has a density of 0.020 kg/m³.
   a A hydrogen balloon floats in air, but would it float in the Martian atmosphere? Explain your answer.
   b Could a balloon filled with air float in the Martian atmosphere? Explain your answer.

CHALLENGE YOURSELF

Make a short presentation called 'The density of matter'.

Decide how you will structure your presentation. How will you build on this structure by providing information that you think is key for the topic? How will you add examples to give interest to the key information points? What will you use to illustrate your presentation to convey key information and interest?

Create your presentation, let others watch the presentation and ask for their comments on which parts work well and which parts could be made clearer.

CHALLENGE YOURSELF

Groupwork

Why do hot-air balloons rise into cold air? Share your ideas as a group or with your teacher. Research the internet to find out how to make a hot-air balloon using a hair dryer.

Write a plan of how you would make the hot-air balloon rise and include a risk assessment in your plan. Show the plan to your teacher and, if approved, try it.

▲ Figure 8.9 A hot-air balloon in flight.

Science in context

Balloons and scientific research

Joseph-Michel Montgolfier (1740–1810) had a paper-making business, but he was also interested in inventions. One day he watched some clothing being dried over a fire and saw how parts of the clothes rose up into the air and then fell back again as the hot air circulated around them. He also saw that the embers from the fire rose into the air and fell back down again, but the smoke kept rising. He came up with the hypothesis that the smoke rose because it contained a property called **levity**, which he believed made things rise against gravity. He believed the smoke contained a gas with the property of levity which made the smoke rise, and he called this gas 'Montgolfier gas'.

▲ **Figure 8.10** Meteorologists use weather balloons to send weather recording equipment high into the atmosphere.

Joseph-Michel made a lightweight wooden box, covered it in a material containing silk and lit a fire under it. The box rose into the air. This experiment led him to ask his brother Jacques-Étienne Montgolfier to join him in more experiments, which led to the launching of a huge balloon made of hessian and paper over a smoky fire.

More balloon flights followed. There was some concern over how living things might survive higher in the atmosphere, so the first living things that were sent up in a balloon were a sheep, a duck and a chicken.

An Italian scientist, Tiberius Cavallo, had made experiments on bubbles by filling them with hydrogen and observing how they rose in the air. This information about hydrogen rising in air was known by the French inventor Jacques Alexandre César Charles, and he invented a hydrogen balloon that could carry people.

Gas-filled balloons were easier to control than the hot-air balloon, and soon became used for other purposes than simply transporting people. Today, the gas used in balloons is helium rather than hydrogen. These balloons are used in scientific research to measure weather conditions high in the air and they provide data for predicting the weather on the ground. They are also used for investigating the atmosphere high above the planet's surface and for astronomy.

12 What observations led Joseph-Michel Montgolfier to set up his hypothesis?

13 A sheep is about the size of a human and does not fly, ducks are used to flying high in the sky, but chickens normally stay only on the ground. How do you think these different animals helped in testing the effects of living things travelling high into the atmosphere?

14 If you were asked to provide the Montgolfier brothers with a risk assessment for their balloon and animal experiment, what would you write?

 15 From your knowledge of chemistry, why do you think helium is considered a safer gas to use in balloons than hydrogen?

 16 From your knowledge of weather-recording equipment, what instruments do you think weather balloons may carry?

LET'S TALK

The Montgolfier experiments used animals in tests to see how living things would survive higher in the atmosphere. Do you think this was a good decision or should humans have gone up instead? Share your ideas.

Summary

✔ The density of a substance depends on the mass of a certain volume.
✔ A formula can be used to calculate densities.
✔ The density of a regular shaped solid can be found by measuring its mass and calculating its volume.
✔ The density of an irregularly shaped solid can be found by calculating its volume by immersing it in water and measuring its mass.
✔ The density of a liquid can be calculated by using a measuring cylinder to find its mass and volume.
✔ Floating and sinking can be explained in terms of density.
✔ The density of a gas can be found by measuring its mass and volume.
✔ Large gas-filled balloons are used to predict weather conditions which can affect societies around the world.

End of chapter questions

1 Arrange the materials in Table 8.1 on page 83 in order of density, starting with the least dense material.

2 Which is heavier – 1m³ of steel or 1m³ of aluminium?

3 Which is heavier – 1kg of steel or 1kg of cork?

4 How can gas density be used to explain why hydrogen rises in air and carbon dioxide sinks?

5 It is claimed that if you could find a lake of water large enough, you could float planet Saturn in it. The density of Saturn is 687 kg/m³. Look at Table 8.2 on page 86 and decide if you agree. Explain how you came to your decision.

 6 If you were writing a science-fiction story and wanted to have buildings floating in the atmosphere of Venus (density of 65 kg/m³), which solid could you use? Look at Table 8.3 on page 90 to help you to decide.

 Now you have completed Chapter 8, you may like to try the Chapter 8 online knowledge test if you are using the Boost eBook.

9 Displacement reactions

In this chapter you will learn:
- to identify displacement reactions
- to use word equations and symbol equations
- how to describe the displacement of hydrogen by magnesium
- about the reactivity series of metals
- how to describe the reaction of metals with acids
- about the wide use of displacement reactions (Science in context)
- about the displacement of metals and how to use it to predict the products of a reaction.

Do you remember?

- Name two acids and two alkalis.
- What is a word equation used for?
- How is a symbol equation different from a word equation?
- In the reactivity series, which of these two metals is the most reactive – gold or sodium?

Displacement reactions

In a **displacement** reaction, a less reactive element in a compound is replaced by a more reactive one. Scientists use the word **displaced** instead of replaced, and this word gives the name to the reaction.

▲ **Figure 9.1** Potassium reacting with water.

The reaction of potassium with water is an example of a vigorous displacement reaction. Water is a compound made from hydrogen and oxygen; in this reaction, the potassium displaces (replaces) the hydrogen, and potassium hydroxide is formed. The reaction takes place so quickly that heat is produced in such large amounts, it sets fire to the hydrogen!

Many other displacement reactions take place much more slowly than this. You are going to investigate reactions involving iron, copper and magnesium, then you will look at reactions involving more metals and consider the reactivity series of metals.

Can iron displace copper from copper sulfate solution?

You will need:

a test tube and test tube rack, copper sulfate solution, a clean iron nail, a camera, a stop-clock.

Hypothesis

If copper is less reactive than iron, it will be displaced from the compound copper sulfate in a copper sulfate solution.

Prediction

If copper is displaced from copper sulfate solution, it will appear as a bright brown metal and the solution will lose its blue colour.

Process

Read through the following steps, then make a table in which to record your data.

1. Look at the clean iron nail and note and photograph its colour. Look at the copper sulfate solution and photograph its colour.
2. Look at the length of the nail, assess its length and then use your assessment to add copper sulfate to the test tube so that it will reach only half-way up the nail.
3. Place the nail in the test tube so that equal lengths are in the solution and in the air.
4. Record the time that you placed the nail in the solution and note the colour of the nail and the solution.
5. Repeat step 4 every 5 minutes for 45 minutes.

Examining the results

Look through your data and see if you can find a pattern.

Conclusion

Compare the examination of your data with the hypothesis and prediction and draw a conclusion.

Is your conclusion limited in some way? Explain your answer.

What improvements could be made? Explain the changes that you suggest.

How was the enquiry made into a fair test?

The word equation for the reaction between iron and copper sulfate is:

iron + copper sulfate → iron sulfate + copper

The symbol equation for this is:

$Fe + CuSO_4 \rightarrow Cu + FeSO_4$

Water is used to check the reactivity of metals. In prior learning, you may have investigated the reaction of zinc, magnesium, iron and copper with water. Only magnesium shows a reaction with water.

The displacement of hydrogen by magnesium

The magnesium slowly displaces hydrogen in the water molecules and forms magnesium hydroxide, releasing hydrogen. The presence of magnesium hydroxide is detected by the indicator phenolphthalein. Phenolphthalein is an indicator which is colourless in almost neutral solutions, but turns pink in weak alkaline solutions and orange in strong acidic solutions. In Figure 9.2, it has turned pink to show the presence of magnesium hydroxide.

▲ **Figure 9.2** Magnesium hydroxide in phenolphthalein. The tube on the right shows magnesium pieces shortly after they were added to the indicator. The tube on the left shows a change in colour in the indicator due to the production of magnesium hydroxide in the chemical reaction.

The word equation for the reaction between magnesium and water is:

 magnesium + water → magnesium hydroxide + hydrogen

The symbol equation is:

 $Mg + H_2O \rightarrow MgOH_2 + H_2$

1 What kind of solution does magnesium hydroxide make?
2 The magnesium in the magnesium hydroxide compound is a cation (a positive ion) and the two OH groups attached to it are anions (negative ions). Which part of the compound lost electrons and which part gained them? (See Chapter 7 if you want to remind yourself about ionic bonds.)
3 The hydrogen molecules produced in this reaction are formed by a covalent bond between the two hydrogen atoms. Explain how the electrons have made this bond. (See Chapter 7 if you want to remind yourself about covalent bonds.)

▼ **Table 9.1** Some metals in the reactivity series, from most to least reactive.

Metal	Symbol
potassium	K
sodium	Na
calcium	Ca
magnesium	Mg
zinc	Z
iron	Fe
copper	Cu
silver	Ag
gold	Au

The reactivity series of metals

Scientists sort out metals into a reactivity series. The table on the left shows some of the metals in the series, starting with the most reactive and ending with the least reactive.

We know that magnesium displaces hydrogen from water. The metals around magnesium (calcium, zinc and iron) also displace hydrogen from water, but the last three (copper, silver and gold) do not displace it. The reactivity series provides a clue to how metals will be displaced when chemical reactions take place between them.

4   Name the metals around magnesium which are
    a   more reactive than magnesium
    b   less reactive than magnesium.

The reaction of metals with acids

When metals react with acids, they displace hydrogen from the acid and form a salt solution. You will find out more about this in Chapter 10. The general word equation for this reaction is:

metal + acid → metal salt + hydrogen

5   Use the general word equation to help you write the word equations for the reactions between
    a   magnesium and sulfuric acid
    b   zinc and hydrochloric acid
    c   calcium and nitric acid.

    (**Hint:** These names might help you with the name of the metal salt – chloride, nitrate, sulfate.)

How do magnesium and zinc react with an acid?

You will need:

safety glasses, a test tube rack, test tubes, boiling tubes, splints, a means of lighting the splints, magnesium, zinc, dilute hydrochloric acid.

Hypothesis

Construct a testable hypothesis using your knowledge about the reactivity of magnesium and zinc with acids.

Prediction

When zinc and magnesium are each added to an acid in turn, what will you observe?

CHALLENGE YOURSELF

Expand on this enquiry by filtering the substances using a clamp and stand, filter paper, filter funnel and a beaker to produce a filtrate. Make a diagram showing how you will set up the equipment and describe how you will use it. Don't forget to wear safety goggles when filtering a liquid that may contain an acid. If your teacher approves, try it. Pour your filtrate into a second boiling tube and carefully seal it with a stopper or bung under the supervision of your teacher, then prepare labels and stick them on as appropriate.

DID YOU KNOW?

The rocks in which metals such as iron are naturally found are referred to as ores.

Planning, investigating and recording data

Make a step-by-step plan using the items listed on the previous page, or any others that you think may be useful. Include your method of recording. Check your plan with your teacher and, if approved, try it.

Examining the results

Make comparisons with items in your data.

Conclusion

Compare the examination of your data with the hypothesis and prediction and draw a conclusion.

Is your conclusion limited in some way? Explain your answer.

What improvements could be made? Explain the changes that you suggest.

How was the enquiry made into a fair test?

Science in context

The wide use of displacement reactions

If you travel by train, you may see a railway line being repaired as you pass by. This repair often involves joining two metal rails together. The rails have to carry the heavy weight of a train, so they need to be firmly connected together. This is done on the rail track by heating the two ends of the rails that are to be joined together, then adding a powder containing aluminium and iron oxide. The heat causes a displacement reaction which produces even more heat.

▲ **Figure 9.3** A displacement reaction is used to weld railway tracks.

In the reaction, aluminium displaces iron from its oxide. The iron melts and joins the melted iron on the ends of the rails, linking them together. When the rails cool, a very strong joint of fused metal is produced, and the track is repaired.

There are two non-metals that can be used in displacement reactions. The first is carbon, which fits between magnesium and zinc in the reactivity series in Table 9.1. This means it can take part in displacement reactions with zinc, iron and copper, and is used to extract these metals from their ores. Items in daily use made from these metals have been produced by displacement reactions.

The second non-metal is hydrogen, which fits between iron and copper in Table 9.1. A displacement reaction takes place involving hydrogen when you treat indigestion with a tablet. The indigestion is caused by excess hydrochloric acid in the stomach and the tablet contains sodium bicarbonate. In the reaction, the more reactive sodium displaces hydrogen to form sodium chloride, and the acidity is removed. You will have met this reaction before when you studied neutralisation in grade 7.

▲ **Figure 9.4** Copper wire coils in silver sulfate solution. Silver is formed on the wire.

▲ **Figure 9.5** An iron nail in copper sulfate solution. Copper has formed on the nail.

Displacement of metals

When metals react with acids, they **displace** hydrogen from the acid and form a salt solution. In a similar way, a more reactive metal can displace a less reactive metal from a salt solution of the metal.

When a copper wire is suspended in a solution of silver sulfate, the copper reacts with the silver sulfate and dissolves into the solution to form copper sulfate. Silver metal comes out of the solution, and settles on the wire, as shown in Figure 9.4.

The word equation for this reaction is:

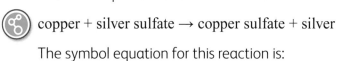 copper + silver sulfate → copper sulfate + silver

The symbol equation for this reaction is:

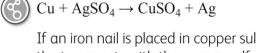 $Cu + AgSO_4 \rightarrow CuSO_4 + Ag$

If an iron nail is placed in copper sulfate solution, the iron reacts with the copper sulfate and dissolves into the solution. This forms a pale green iron sulfate solution. The copper comes out of the solution and coats the nail, as shown in Figure 9.5.

The word equation for this reaction is:

 iron + copper sulfate → iron sulfate + copper

The symbol equation for this reaction is:

 $Fe + CuSO_4 → Cu + FeSO_4$

6 In an investigation into displacement reactions, make predictions and explain your answers for reactions between
   a silver metal and gold nitrate
   b zinc in a copper sulfate solution
   c magnesium with copper sulfate
   d gold with magnesium chloride.

Summary

✔ In a displacement reaction, a less reactive element is replaced by a more reactive element.
✔ Word and symbol equations can be used to show displacement reactions.
✔ Magnesium replaces the hydrogen in water molecules and forms magnesium hydroxide.
✔ The reactivity series of metals can be used to predict which displacement reactions will occur.
✔ The reaction of metals with acids can be summarised by the word equation:

 metal + acid → metal salt + hydrogen

✔ Displacement reactions are used to repair railway tracks around the world.

End of chapter questions

1 Describe what happens when a piece of potassium is added to water.
2 Explain the chemical reaction that takes place when a piece of potassium is added to water.
3 Why does a chemical reaction take place when an iron nail is placed in copper sulfate solution?
4 Why should there be a difference in the reaction between zinc and acid, and magnesium and acid?

 Now you have completed Chapter 9, you may like to try the Chapter 9 online knowledge test if you are using the Boost eBook.

10 Preparing common salts

In this chapter you will learn:
- about acids and their salts
- how to use word equations to describe reactions
- about purification through evaporation, crystallisation and filtration
- how to prepare a salt from a metal and an acid
- how to prepare a salt from a metal carbonate and an acid
- about Salimuzzaman Siddiqui and chemicals from natural products (Science in context).

Do you remember?

- Where have you come across the word 'salt' in chemistry before?
- What is an acid?
- Look at the pH scale in Figure 10.1. Which numbers denote substances that are acids?
- What is the name of the indicator that is used to measure the pH of an acid or an alkali?
- How do metals react with acids?
- What is a 'pure' substance?
- What is evaporation?

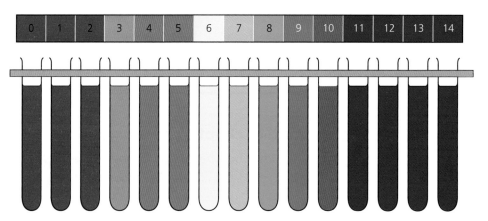

▲ **Figure 10.1** The pH scale.

▲ **Figure 10.2** Table salt. Its chemical name is sodium chloride.

▲ **Figure 10.3** Zinc sulfate is a metal salt that has many uses in industry, including in the manufacture of rayon fibres from cellulose.

Salts

Some people have difficulty thinking about the word 'salt'. All they can think about is sodium chloride – common salt – as shown in Figure 10.2. However, there are many different salts, and some of them have a wide range of uses.

Calcium chloride is a salt that forms white crystals. It is used to absorb moisture from the air and is known as a drying agent. It is also used in food processing and seasoning, in medicine and in speeding up the setting of concrete.

Zinc sulfate also forms white crystals. Some of its uses include making cosmetics, the textile material rayon, glue and in the bleaching of paper (bleaching is the process of making paper white so that it is easier to print on), as well as a **herbicide** and in the treatment of sewage. It is also used by chemists in investigations to find the chemical components of different substances.

Acids and their salts

This chapter will explore two methods of preparing salts – using metals and acids, and using metal carbonates and acids. The word 'carbonate' means that the chemical compound contains carbon and oxygen linked together. A metal carbonate, such as calcium carbonate, is a compound that contains calcium, carbon and oxygen. The chemical reactions that take place in these two methods of preparation are represented by these general word equations:

 acid + metal → salt + hydrogen

 acid + carbonate → salt + carbon dioxide + water

The three acids that can be used to make salts are hydrochloric acid, sulfuric acid and nitric acid. The salts produced by these acids are chlorides, sulfates and nitrates.

▲ **Figure 10.4** Some examples of chloride, sulfate and nitrate salts.

Purification by evaporation, crystallisation and filtration
Evaporation and crystallisation

▲ **Figure 10.5** Crystals of copper sulfate in an evaporating dish.

A crystal is a solid structure with flat sides. Many substances form crystals.

One way of making crystals in the preparation of salts is to start with a concentrated solution of a substance. When the concentrated solution is gently heated in an evaporating dish, the solvent begins to evaporate. This causes the solution to become even more concentrated.

If the heat is removed and the concentrated solution is left to cool, evaporation will continue even when it reaches the temperature in the laboratory. In time, all the solvent will evaporate and pure crystals of the substance will be left behind.

The pure salt crystals can then be dried by patting them gently and carefully with a paper towel. When salts are being prepared, this is the last part of the procedure.

Filtration

In many laboratory experiments, filtration is carried out by folding a piece of filter paper to make a cone and inserting it into a filter funnel. The funnel is then supported above a collecting vessel and the mixture to be separated is poured into the funnel (see Figure 10.7 on the next page).

The filter paper is made of a mesh of fibres. It works like a sieve but the holes between the fibres are so small that only liquid can pass through them. The solid particles are left behind on the paper fibres.

The pure substance left behind on the filter paper is called the residue, and the liquid that passes through the filter paper is called the filtrate.

> **DID YOU KNOW?**
> A highly concentrated solution is called a **saturated solution**. It is a solution that is so packed with a solute that no more will dissolve in it.

Preparing a salt from a metal and an acid
Preparation of zinc chloride

Small fragments of zinc are added to hydrochloric acid in a flask (see Figure 10.6 on the next page). Bubbles of gas rise from the metal, pass through the liquid and escape into the air. Eventually, the bubbles are no longer produced and some metal remains in the flask.

The contents of the flask are then poured onto filter paper in a filter funnel (see Figure 10.7 on the next page). The zinc metal remains behind and the liquid passes through and falls into a beaker.

The liquid is then poured into an evaporating dish and heated gently until some solid appears (see Figure 10.8). The mixture is then left to cool, and more evaporation takes place until only the crystals are left behind.

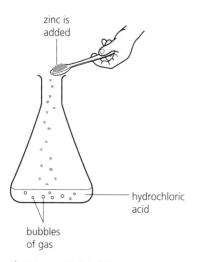

zinc is added

hydrochloric acid

bubbles of gas

▲ **Figure 10.6** Adding zinc to hydrochloric acid.

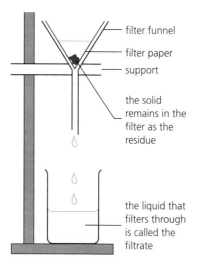

filter funnel

filter paper

support

the solid remains in the filter as the residue

the liquid that filters through is called the filtrate

▲ **Figure 10.7** Filtration with a filter funnel.

▲ **Figure 10.8** Evaporation.

1 Why do you think that granulated zinc is used instead of a block?
2 What passed through the filter paper when the flask was emptied?
3 Why was the solution heated before it was left in an evaporating dish?
 4 Write the word equation for this reaction.
 5 Here is the symbol equation for a reaction:

$$Zn + 2HCl \rightarrow ZnCl_2 + H_2$$

What does this equation tell you about the reaction?
 6 Zinc sulfate can be prepared in a similar way. Write a word equation for the reaction.
7 Here is the symbol equation for a reaction:

$$Zn + H_2SO_4 \rightarrow ZnSO_4 + H_2$$

What does this equation tell you about the reaction?
 8 Write the word equations for the reactions between the following metals and acids.
  a magnesium and sulfuric acid
  b iron and nitric acid
  c calcium and hydrochloric acid
  d lead and sulfuric acid
  e aluminium and hydrochloric acid
  f tin and nitric acid

Can you prepare magnesium sulfate and zinc sulfate salts?

Using all the information in this section, write down:

1 a step-by-step plan to prepare a salt
2 a list of everything you will require, including eye protection
3 what you will do to make sure you work safely.

Check your plans with your teacher and, if approved, move on to the next stage.

4 Gather your equipment and materials and use them with your plan to make a preparation of your salt.
5 Display the salts that you have prepared and, if possible, photograph them.

CHALLENGE YOURSELF

Take the labelled test tubes or boiling tubes from the enquiry titled 'How do magnesium and zinc react with an acid?' in the last chapter and make a plan to produce solid salts from their solutions. If your teacher approves, try it and then display your salts.

Preparing a salt from a metal carbonate and an acid

Preparation of calcium chloride

The carbonate used in this reaction is calcium carbonate in the form of marble chips. Some marble chips are added to hydrochloric acid in a flask. Bubbles are produced and the chips dissolve. Some more chips are added, and more bubbles are produced and then the reaction stops and some of the chips are left in the solution.

Then the contents of the flask are poured onto a filter paper in a filter funnel and the solution and chips are separated (see Figure 10.9). The liquid is poured into an evaporating dish and heated until some solid appears. The mixture is then left to cool, and more evaporation takes place. When the mixture has been left to cool, it is filtered again.

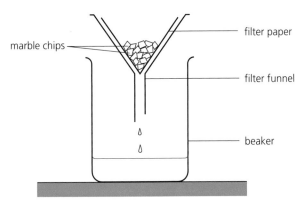

▲ **Figure 10.9** Filtering out marble chips using a filter funnel.

9 Why did all the chips added at first dissolve?

10 Why did some of the chips added later not dissolve?

11 Why were the contents of the flask filtered?

12 Write the word equation for the reaction.

13 a Write the word equations for the reactions between the following metal carbonates and acids:
   i zinc carbonate and sulfuric acid
   ii aluminium carbonate and hydrochloric acid
   iii magnesium carbonate and nitric acid
   iv copper carbonate and sulfuric acid
   v calcium carbonate and nitric acid
   vi lead carbonate and hydrochloric acid

  b Here are three symbol equations for the reactions above. Match them with their word equations.
   i $Al_2(CO_3)_3 + 6HCl \rightarrow 2AlCl_3 + 3CO_2 + 3H_2O$
   ii $CaCO_3 + 2HNO_3 \rightarrow Ca(NO_3)_2 + CO_2 + H_2O$
   iii $ZnCO_3 + H_2SO_4 \rightarrow ZnSO_4 + CO_2 + H_2O$

Salt preparation project

▲ **Figure 10.10** Teamwork.

Scientists often work together in groups on a project. Your group project is to make salts using the following compounds:
- magnesium carbonate
- calcium carbonate
- copper carbonate.

Your target is to make samples of the following:
- magnesium chloride
- magnesium sulfate
- magnesium nitrate
- calcium chloride
- calcium nitrate
- copper chloride
- copper sulfate
- copper nitrate.

(**Note:** When calcium carbonate reacts with sulfuric acid, an insoluble layer of calcium sulfate forms which stops the reaction, so this salt is excluded from the project.)

Look back at page 101 to see which acids are used to prepare salts.

Can your lab team complete a salt preparation?

You will need:

conical flasks, spatulas, evaporating dishes, filter funnels, filter papers, tripods, gauzes, Bunsen burners and heat-proof mats, eye protection.

Planning and investigating

As a group, work out a plan to make several salts at the same time. Write down all the things you need to do to control any risks. If your teacher approves, try it.

Examining the results

At the end of your enquiry, display your salts.

LET'S TALK

- As a group, discuss which parts of the salt preparation project went well and which parts could have been done better.
- How well did people feel they worked together? Did the group collaborate and communicate well?
- Was there another way that other groups worked – perhaps in sub-teams? If your group was to be set another project, would you want to change the way you worked together? Explain your answer.
- If the team activity was to be repeated and filmed for a science fair, how would you organise the team? Would some group members be moved to different roles, such as the commentator, or camera or sound operator?

Salimuzzaman Siddiqui and chemicals from natural products

Chemists all over the world prepare salts on a daily basis. Some of this work is directed at extracting chemicals from plants for use in medicines and **insecticides**. One of the pioneers of this work was Salimuzzaman Siddiqui (1897–1994) who was born in Subeha, India. He attended school in Lucknow, went on to university in Aligarh and later moved to London to study medicine, but after a year he moved to Frankfurt in Germany to study chemistry.

▲ **Figure 10.11** Snake root (*Rauwolfia serpentina*).

Siddiqui returned to India and, guided by Hakim Ajmal Khan (1868–1927), he set up a research institute. A plant known as snake root or sarpagandha (*Rauwolfia serpentina*) has been used for thousands of years in India to treat ailments, and Siddiqui began work on extracting chemicals from its roots. In 1931, he extracted from these roots a substance that cures abnormal beating of the heart. He named it 'ajmaline' in honour of his colleague. Siddiqui went on to discover more chemicals from the plant's roots and many are still used today to treat heart disease and some mental health conditions.

▲ **Figure 10.12** Siddiqui discovered many extremely useful chemical substances in the tissues of plants such as this neem tree.

In 1942, he used a range of solvents, including **ether** and dilute alcohol, to extract chemicals from the neem tree or 'heal-all' (*Azadirachta indica*). These chemicals can be used to kill bacteria, viruses, fungi, **parasitic** worms and can be used as natural insecticides. He continued with his work, discovering over 50 substances, and went on to discover more in research projects with other scientists and their students.

In 1951, Siddiqui was asked to set up research programmes in Pakistan and, in 1953, he founded the Pakistan Academy of Sciences and the Pakistan Council of Scientific and Industrial Research. In 1983, he helped to set up, and became a founding member of, the Third World Academy of Sciences, now known as The World Academy of Sciences (TWAS). Today, TWAS continues Siddiqui's work across a wide range of science topics and has a

14 Would you say that the young Salimuzzaman Siddiqui was interested in the world around him? Explain your answer.

15 What signs of a scientist must he have displayed in his early interests?

16 Why do you think the scientific names of the plants have been included in the text?

17 A scientist should be able to work with others. What evidence can you see of Siddiqui working with others?

18 Would you say that Siddiqui had a life-long passion for science? Explain your answer.

section called TWAS Young Affiliates Network (TYAN), which brings young scientists together from across the world to share and discuss their work.

At the age of 93, Siddiqui retired from running the institute but carried on researching in his laboratory.

▲ **Figure 10.13** Siddiqui working in his laboratory.

LET'S TALK

Imagine you were to follow your studies in science into researching a scientific topic, like the young scientists from the TWAS Young Affiliates Network. What topic would you like to research? Why would you want to do it? What do you think your life might be like as a scientist?

You may find it helpful to spend some time on the internet looking at the TWAS Young Affiliates Network website before or during your discussion: https://tyan.twas.org.

Summary

✔ The preparation of salts can be summarised by these word equations:

acid + metal → salt + hydrogen

acid + carbonate → salt + carbon dioxide + water

✔ The acids used to make salts are hydrochloric acid, sulfuric acid and nitric acid.

✔ Purification occurs through filtration, evaporation and crystallisation in the preparation of salts.

✔ Salimuzzaman Siddiqui was one of the first people to develop a technique for extracting chemicals from plants to prepare salts that are used in medicines and insecticides.

End of chapter questions

1 What is produced when you set up a reaction between a metal and an acid?

2 What is produced when you set up a reaction between a metal carbonate and an acid?

3 How would you filter a mixture of a solid and a liquid?

4 Name two uses of calcium chloride.

5 Name four uses of zinc sulfate.

6 Describe a crystal in one sentence.

7 You have a concentrated solution of copper sulfate. Describe the cystalisation process you would use to make copper sulfate crystals.

8 Name all the equipment you would use to prepare a salt from an acid and a metal.

 9 How has the work of Salimuzzaman Siddiqui helped to improve our scientific knowledge and understanding?

 Now you have completed Chapter 10, you may like to try the Chapter 10 online knowledge test if you are using the Boost eBook.

In this chapter you will learn:
- to use word and symbol equations to describe reactions
- that mass is conserved in chemical reactions
- about closed and non-closed systems in chemical reactions (Science extra)
- that energy is conserved in chemical reactions
- about Einstein's equation (Science extra)
- to measure rates of reactions
- to identify factors that affect the rate of reaction
- about the triangle of fire
- about catalysts (Science in context)
- to use the particle theory to explain rates of reaction.

Do you remember?

- What do you understand by the scientific term 'mass'?
- In chemistry, what do the words 'reactants' and 'products' mean?
- If one liquid is more concentrated than another, what does that mean?
- What do you understand by the word 'temperature'?

Rates of reaction

Chemical reactions take place at different rates. The chemical reaction that blows rock out of a cliff face in a quarry is a very fast reaction. Some reactions, such as the setting of concrete, are very slow and may take up to two days or more to complete.

▲ **Figure 11.1** An explosion in a quarry happens as a result of a very rapid reaction.

▲ **Figure 11.2** When concrete sets, the reaction takes up to two days or longer.

The conservation of mass

A chemical reaction can be summarised in this equation:

 reactant A + reactant B → product C and product D

In every chemical reaction, the mass of matter involved is **conserved**. This means that the mass of products C and D is the same as the mass of reactants A and B. Although the atoms have been rearranged to form different compounds, the mass is neither created nor destroyed. It stays the same – we say that it is conserved.

In some chemical reactions, all the reactants and products stay together. Reactants do not enter the reaction from elsewhere (such as oxygen coming in from the air) and products do not escape from where the reaction has taken place (such as carbon dioxide escaping out into the air). In other words, no new elements combine to make new products.

Science extra: Systems

Chemical reactions where all the reactants and products stay together are called **closed systems**. An example of closed system chemical reactions are the neutralisation reactions of acids and alkalis.

Chemical reactions in which reactant and products can enter and leave the reaction are called **non-closed systems**. Any reactions which involve gases entering or leaving from the air are examples of non-enclosed systems.

The conservation of energy

We know that energy can be stored as chemical energy and this energy can be transferred from one store to another. We represent this through energy transfer diagrams, like this one for the burning of gas in a Bunsen burner:

 chemical energy → Bunsen burner → light energy and heat energy

The Bunsen burner is where the energy transfer takes place.

> **Science extra: Einstein's equation**
> Albert Einstein studied energy transfers and developed one of the most famous equations in science: $E = mc^2$. This means that the movement (or kinetic energy) of an object is the same as its mass multiplied by the speed of light squared. It shows that mass can be changed into energy and energy can be changed into mass, and that during this process, energy is neither created nor destroyed.

Measuring rates of reaction

In chemistry, the rate of a reaction is studied by considering the rate at which the chemicals in the reaction change. Rate is a measure of the change in a certain amount of time. The rate may show how much the mass of the reactants changes in a certain amount of time, or how much product is produced in a certain amount of time.

Change in mass of reactants

The change in mass of the reactants can be investigated using calcium carbonate (in the form of marble chips) and hydrochloric acid.

After the mass of the reactants (marble chips and hydrochloric acid) has been recorded, the reactants are mixed together in the flask and their total mass is recorded, as shown in Figure 11.3 on the next page. The mass is then recorded regularly over a certain number of minutes. The word equation for the reaction is:

 calcium carbonate + hydrochloric acid → calcium chloride + carbon dioxide + water

The symbol equation for this reaction is:

 $CaCO_3 + HCl \rightarrow CaCl_2 + CO_2 + H_2O$

The carbon dioxide escapes from the flask and accounts for the change in mass.

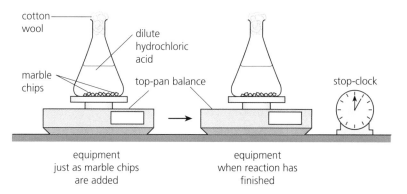

▲ **Figure 11.3** Finding the mass of reactants before, during and after the reaction.

What happens to a mass of marble chips when hydrochloric acid is added?

Hypothesis
Using the information in this chapter, construct a testable hypothesis to explain what will happen to a mass of marble chips when hydrochloric acid is added to it.

Prediction
Construct a prediction from your hypothesis which includes how long it will take for the reaction to be completed.

Planning, investigating and recording data
1 List all the items that you will need. Use Figure 11.3 to help you.
2 Construct a table in which to record your data.
3 Work out a sequence for placing the reactants in the flask so that their mass can be quickly measured at the start of the reaction.
4 Explain how you will know when the reaction has finished and what time you should measure the mass of the reactants again.
5 Check your plan with your teacher and, if approved, try it.

Examining the results
Look at the data you have collected to see if there is a change in the mass.

Conclusion
Compare your examination of the results with your hypothesis, prediction and the title of the enquiry, and draw a conclusion.

Is your conclusion limited in some way? Explain your answer.

What improvements could be made? Explain the changes that you suggest.

Change in volume of a product

In Figure 11.4 when the reactants (magnesium and hydrochloric acid) are mixed together, hydrogen is produced, as the following word equation describes:

 magnesium + hydrochloric acid → magnesium chloride + hydrogen

The symbol equation for this reaction is:

 $Mg + 2HCl → MgCl_2 + H_2$

As the gas is produced, it pushes the plunger in the syringe to the right, and the volume produced every minute can be measured.

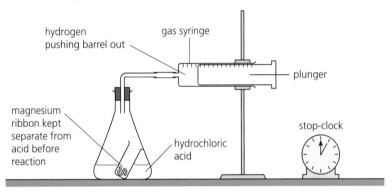

▲ **Figure 11.4** The volume of hydrogen produced can be measured.

Factors affecting rates of reaction

Concentration

The concentration of a solution is a measure of the amount of solute dissolved in it. A solution of low concentration has a small amount of solute dissolved in it. A solution of high concentration has a large amount of solute dissolved in it. If the concentration of a reactant is increased, the rate of reaction increases. If the concentration of one reactant is doubled, the rate of the reaction may be doubled.

Sometimes, scientists look back at experiments and activities they have done and develop them into new ones. A model volcano can be examined as an example of a neutralisation reaction. You may have seen or made a model volcano in earlier science studies.

▲ **Figure 11.5** The ingredients (left) for making a model volcano (right).

In the model volcano, some vinegar is poured onto sodium hydrogen carbonate. A reaction takes place, producing a fizzy liquid that flows down the sides of the volcano like lava. This activity can be looked at again in a new way to investigate the effect of the concentration of vinegar on the rate of reaction.

> **How does the concentration of vinegar affect the rate of reaction in a model volcano?**

You will need:

vinegar, sodium hydrogen carbonate, water, a bottle, a filter funnel, a tray, a stop-clock or timer, a measuring cylinder, a top-pan balance or other weighing machine.

Hypothesis

Construct a testable hypothesis to explain what will happen if different concentrations of vinegar are reacted with the same amount of sodium hydrogen carbonate.

Prediction

Construct a prediction based on your hypothesis.

Process

1 Measure out a certain volume of vinegar with the measuring cylinder and then pour it into the bottle.
2 Weigh out a certain amount of sodium hydrogen carbonate. Insert a filter funnel into the top of the bottle and pour the powder into it so

that all of it goes into the vinegar in the bottle. Remove the funnel immediately and start the clock or timer.

3 Stop the clock or timer when all the fizzing has stopped and record the time.

4 Make a more dilute solution of the vinegar and repeat steps 2 and 3.

Examining the results

Look at the data for signs of a trend or pattern. Do any results not fit the pattern? Can you explain why any results might be anomalous?

Conclusion

Compare your examination of the results with the hypothesis and prediction and draw a conclusion.

Is your conclusion limited in some way? Explain your answer.

What improvements could be made? Explain the changes that you suggest.

CHALLENGE YOURSELF

Groupwork

Scientists often try to devise alternative experiments to test a hypothesis and prediction. They use other knowledge to construct different experiments. Earlier, for example, you saw how a syringe could be used to measure the volume of a gas.

Using only an empty bottle, filter paper, vinegar, baloons and sodium hydrogen carbonate, how do you think you could adapt the enquiry about the model volcano? Work with a partner, first construct a step-by-step plan for the experiment and, if your teacher approves, try it.

Do your results support the evidence collected in the model volcano enquiry? Explain your answer.

Particle size

When a piece of coal is heated, it produces a flame and burns steadily in air. When coal dust is heated, it explodes. The reason for this difference in reaction rate is due to the surface area of coal in contact with oxygen in the air. When a piece of coal is ground into dust, it has a much larger surface area in contact with the oxygen in the air. This suggests that particle size affects the rate of reaction.

▲ **Figure 11.6** Pieces of burning coal.

▲ **Figure 11.7** Tests on coal dust at Bełchatów Power Station result in exploding dust.

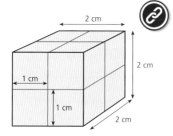

▲ **Figure 11.8** One large cube can be cut into eight smaller cubes.

When a solid, such as a piece of coal, breaks up into smaller particles, its surface area increases, as the following example shows. Imagine a cube-shaped piece of coal with sides 2 cm long. It has six surfaces. Each one is $2 \times 2 = 4\,cm^2$ in area. The surface area of the cube is $6 \times 4 = 24\,cm^2$.

If the cube is broken into eight cubes, each with a side of 1 cm, the surface area of each cube would be $6 \times 1 \times 1 = 6\,cm^2$. As there are now eight cubes, their total surface area is $8 \times 6 = 48\,cm^2$. The surface area has doubled. If the eight cubes were broken up into smaller pieces, the surface area would increase even more.

The surface area of a reactant is its point of contact with other reactants. If the surface area is increased, the reactants can come into contact more rapidly and the reaction rate will increase.

1 The graph in Figure 11.9 shows the volume of carbon dioxide produced when large marble chips take part in a reaction with hydrochloric acid.
   a Make a copy of the graph and draw in the line you would expect to see when smaller chips are used.
   b Explain your answer.

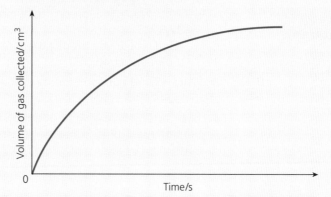

▲ **Figure 11.9** Volume of carbon dioxide produced over time.

Temperature

The rate of reaction increases if the temperature is raised. The rate of reaction decreases if the temperature is lowered. If the temperature of the reactants is raised by 10 °C, the speed of the reaction may be doubled.

Sodium thiosulfate is a substance that dissolves in water. When hydrochloric acid is added to a solution of sodium thiosulfate, sodium chloride, water, sulfur and sulfur dioxide are produced. The sulfur forms a **precipitate** that clouds the solution and the speed at which this cloudiness appears can be used as a measure of the rate of reaction. An investigation can be carried out in the following way, as shown in Figure 11.10.

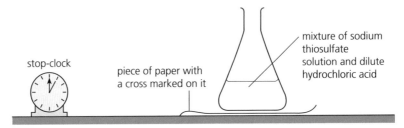

▲ **Figure 11.10** An experiment to assess how temperature alters rate of reaction.

1 A flask containing the reactants is placed over a piece of paper with a cross on it.
2 The reactants are viewed from the top of the flask and a stop-clock or timer is started.
3 When the sulfur precipitate clouds the solution so much that the cross can no longer be seen, the stop-clock or timer is stopped and the time is recorded.

If the experiment is repeated several times, with the reactants at increasingly higher temperatures, a table of data can be produced (see Table 11.1).

▼ **Table 11.1** Data table of temperature and time.

Temperature/°C	Time of reaction/s
25	110
30	80
35	60
40	46
45	38
50	30

2 a Plot the data in the table as a graph.
 b What is the shape of the graph?
 c What would you predict the time of the reaction to be at 32.5 °C?
 d What would you predict the temperature to be if the reaction took 76 seconds?

3 Write a word equation for the reaction between sodium thiosulfate and hydrochloric acid.

4 Two cartons of milk were opened; one was left by a radiator while the other was placed in a fridge. How will the milk in the two cartons differ after three days? Explain your answer.

▲ **Figure 11.11** The triangle of fire.

The triangle of fire

A model called the triangle of fire has been developed to help people understand the key features which cause a fire and how removing any one of them can control the fire and put it out.

Fire is one of the most harmful chemical reactions. It can occur in homes, schools, factories or other buildings, and can occur naturally or purposefully in ecosystems such as forests, as shown in Figure 11.12 below.

If a fire occurs on a rubbish tip, or on a pile of wood, the rubbish or the wood is the **fuel**. This fuel can be removed from around the fire by firefighters, so that there is no more left to burn. See Figure 11.13 below.

▲ **Figure 11.12** Forest fire.

▲ **Figure 11.13** Firefighters tending to a fire at a rubbish dump in Greece.

The heat from the fire can be reduced by pouring water on it, and oxygen can be removed by covering the fire in foam, but there are safety issues. Water should not be poured on burning oil or petrol as it will quickly boil and spread the fire to other places. A common cause of fire in a home is a burning pan of oil. This can be extinguished by covering the top of the pan with a fire blanket to prevent oxygen from reaching the fire. Water and foam should not be used on burning electrical appliances as they conduct electricity and could give firefighters a shock.

5 We have seen that the triangle of fire has limited use when fighting forest fires. Could water still be useful in fighting a forest fire? Explain your answer.

The triangle of fire is a useful model to remember when dealing with small fires, but it is of limited use to firefighters who are dealing with forest fires. There is no way that they can cut off the supply of oxygen in the air around the burning trees, so they have to set up fire breaks between groups of trees in which any material (wood) is removed between the groups to stop the fire from spreading further.

Catalysts

It was known from ancient times that some substances could be added to a chemical reaction, such as the fermentation of wine, which would speed it up. In 1540, a German scientist called Valerius Cordus (1515–44) used sulfuric acid to speed up a reaction which converted alcohol into ether.

Elizabeth Fulhame lived during the eighteenth century, and studied chemical reactions on metals and water and other solvents. One of her discoveries was that the addition of some substances to reactants made them react faster. In 1794, she wrote a book in which she clearly described the equipment, materials (chemicals) and the methods she used in her research so that others could try them, review them and develop them further.

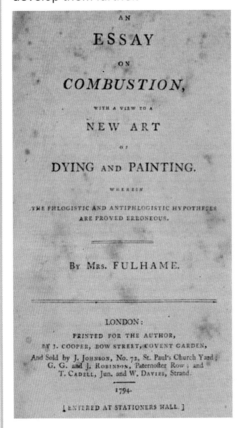

▲ **Figure 11.14** Fulhame lived before the time of photography, but shown here is a photograph of Fulhame's book.

Fulhame's work was widely read by other scientists over the years, and led to many developments, as she had hoped. Scientists such as Humphrey Davy (1778–1829), who discovered several elements, and Johann Wolfgang Döbereiner, who spent time sorting out the elements, perfomed experiments in which substances were added to reactions to speed them up.

In 1835, Jöns Jakob Berzelius (1779–1848) devised the word **catalyst** to describe these substances. Today we use the word to describe a substance that is added to reactants to change their rate of reaction but does not change itself. Usually, catalysts are used to speed up reactions.

Catalysts are used in converting the wide range of molecules found in oil into molecules that can be used to make plastics; they are also used to speed up chemical reactions to make fertilisers, and in the production of acids, which have many uses in industry.

Catalysts are also used in vehicles to make **catalytic converters**. A catalytic converter receives harmful waste gases, such as nitrous oxide and carbon monoxide from a car engine and quickly converts it into water, nitrogen and carbon dioxide. In the past, these gases were simply released into the air through the car's exhaust pipe. In cities, these gases gathered to form a **photochemical smog** which damages the respiratory systems of humans and animals.

▲ **Figure 11.15** A catalytic converter.

▲ **Figure 11.16** Photochemical smog in an urban area.

The use of catalytic converters brings about a chemical change in the exhaust gases so that water, nitrogen and carbon dioxide are produced, and air pollution is reduced. As carbon dioxide is now identified as a **greenhouse gas** which is bringing about climate change, vehicles are being developed to run on electricity which does not produce any air pollution at all.

DID YOU KNOW?

There are catalysts at work in your body at this very moment. They are called **enzymes** and they work in every cell of your body, speeding up the chemical reactions that take place to keep you alive.

6 What features of Fulhame's work helped others who followed her in the study of catalysts?

7 'Scientists are interested in only one topic.' Do you agree? Use the information here to explain your answer.

8 Are catalysts important in today's world? Explain your answer.

The particle model and rates of reaction

We have seen that the particle model of matter can be used to explain physical changes such as melting and freezing. The particle model of matter can also be used to explain the factors that affect the rates of reactions. Particles take part in reactions when they collide together, so any factors that increase the chance of collisions will increase the rate of reaction.

Concentration

A concentrated solution has more particles that are available to react in it than a dilute solution does. This means that increasing the concentration of a solution increases the number of particles, and thus increases the number of collisions and the rate of reaction.

Surface area

On page 117, we saw how a large cube had a smaller surface area than many smaller ones. Reactions take place at surfaces – so the smaller the surface area, the smaller the chance of collisions between the particles in the surface and the particles of the reactant in the liquid or gas next to the surface. Increasing the surface area increases the chance of collisions and so increases the rate of reaction.

Temperature

The speed at which particles move depends on their temperature. If the temperature is raised, the speed of the particles increases and they make harder collisions, which are more likely to result in reactions. The increased speed of the particles also increases the chance of collisions taking place.

Summary

✔ Mass and energy are conserved in chemical reactions.
✔ The rate of reaction is a measure of the rate at which the chemicals in a reaction change.
✔ Rates of reactions can be found by measuring changes in the mass of the reactants.
✔ Rates of reactions can be found by measuring changes in the volume of a product.
✔ The rate of reaction is affected by the concentration, particle size and temperature of the reactants.
✔ Word and symbol equations can be used to describe reactions.
✔ The particle model can be used to explain rates of reactions.
✔ Scientific understanding about the role of catalysts was developed by many individuals over time, and they are now widely used in industry.

End of chapter questions

1 What does the word 'rate' mean when measuring reactions?

2 What would you expect to happen to a mass of calcium carbonate in marble chips when placed in an acid? Explain your answer.

3 State two ways you could collect gas from a chemical reaction.

4 a If you had a 1cm³ cube of a substance, what is its surface area?

   b If you cut the 1cm³ of a substance to 1mm cubes, how many tiny cubes would you have?

   c What is the surface area of all the tiny cubes?

   d Compare the surface area of the 1cm³ cube with the surface area of all the tiny cubes. What do you find?

5 What happens to the rate of reaction between two substances when the temperature is increased? Explain your answer.

6 Shazia has built a campfire and it is burning well. Robert collects some damp logs and puts them on the fire. Shazia is annoyed with Robert because the fire now burns more slowly. Why do you think there has been a change in the rate of reaction?

7 Once the wood has dried out and the fire is burning well again, Robert challenges Shazia to devise a simple experiment using the wood from a dead tree and a bush to show that particle size affects the rate of reaction. What should Shazia do?

Now you have completed Chapter 11, you may like to try the Chapter 11 online knowledge test if you are using the Boost eBook.

In this chapter you will learn:

- about the relationship between thermal energy (heat) and temperature
- that thermal energy can be used to compare different substances (Science extra)
- how energy is conserved
- about heat dissipation and how heat is transferred through conduction, convection and radiation
- about the structure of metals (Science extra)
- how evaporation cools things down
- about the uses of cooling by evaporation (Science in context)
- about thermal imaging (Science in context)
- the reason why greenhouses warm up (Science extra).

Do you remember?

- What is the scientific name for movement energy?
- What type of energy is found in foods?
- What kind of energy do you detect with your ears?
- What energy changes take place when you
  a burn gas in a Bunsen burner
  b blow up a balloon?
- What is wasted energy?
- What do you think the word 'energy' means? Look at Figure 12.1 for some ideas.

▲ **Figure 12.1** Energy.

Thermal energy, internal energy and temperature

Thermal energy is often called heat energy. The 'heat' in a substance is really a measure of the total kinetic (movement) energy of the atoms and molecules of the substance, due to its internal energy. The total amount of heat in a substance is related to its mass.

When a substance is heated, the thermal energy supplied increases the internal kinetic energy. This means the atoms and molecules in the substance move faster and further. If the temperature of the substance is taken with a thermometer, kinetic energy from the substance passes to the atoms or molecules of the thermometer liquid and causes them to move faster too. This leads to an expansion of the liquid in the thermometer tube.

The thermometer measures the (average) kinetic energy of the particles hitting the bulb and not the total kinetic energy of all the particles in the substance.

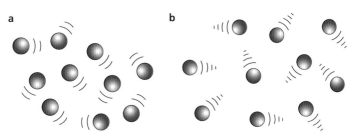

▲ **Figure 12.2** Particles in motion in **a** a cool substance and **b** a hot substance.

Heat is a form of energy and can be transferred from one object to another.

Temperature is an indication of how hot or cold an object is and can be measured with a thermometer. Temperature is a property of an individual object.

Measuring the amount of thermal energy

The amount of thermal energy can be measured by simply using a thermometer.

Look at the images in Figure 12.3. Use them to help you complete the enquiry about thermal energy that follows.

 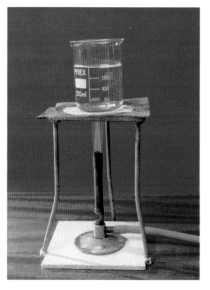

▲ **Figure 12.3** The different volumes of water in the beakers have different masses and are receiving thermal energy.

> Does providing the same amount of thermal energy to different masses of water make them rise to the same temperature?

Hypothesis

Construct a testable hypothesis to provide an answer to the enquiry question.

Prediction

Construct a prediction from your hypothesis.

Planning and investigating

Write down a list of equipment you will need to make your investigation. Then write a step-by-step procedure to test your prediction. If your teacher approves your plan, try it.

Examining the results

Look for patterns and trends in your data.

Conclusion

Compare your evaluation with your hypothesis and prediction and draw a conclusion.

Compare your conclusions with that of other students in your class. Discuss any limitations and ways in which the investigation could be improved.

Science extra: Comparing different substances

The amount of heat (thermal energy) given to a substance can be measured by heating the substance with an electric heater. The quantity of electrical energy used can be measured using a **joulemeter** and this equals the amount of heat (thermal energy) supplied.

The equipment shown in Figure 12.4 on the next page can be used to compare how the heat supplied to a liquid or solid affects its temperature. It is found that some substances, such as water, take up large amounts of heat energy, but their temperature only rises a few degrees, while the same mass of other substances may need only a small amount of heat energy to raise the temperature by the same amount.

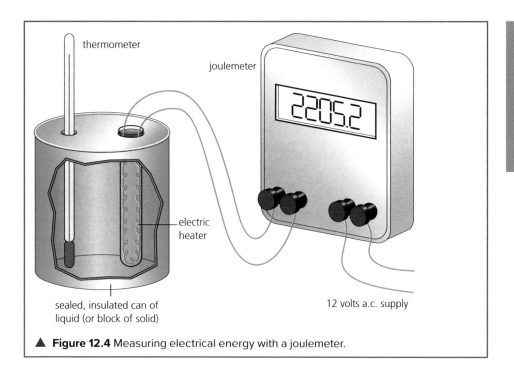

▲ **Figure 12.4** Measuring electrical energy with a joulemeter.

Conservation of energy

Chapter 11 explored how mass is neither created nor destroyed, but is conserved, and how energy is also neither created nor destroyed.

Every form of matter, from an atom to giant molecular and ionic structures, possesses energy. In a chemical reaction, some of this stored energy is released from the reactants as the bonds break between their atoms, and some energy is stored in new bonds made between the atoms of the products.

In exothermic reactions, energy is released as heat and sometimes also as light into the surroundings. In **endothermic reactions**, energy is taken from substances in the surroundings and stored in the products. No energy is destroyed or created; it simply moves from one form to another and is conserved.

3 How is energy being conserved as the cheetah starts to move?

▲ **Figure 12.5 a** A cheetah at rest and **b** the cheetah on the move.

▲ **Figure 12.6 a** A young plant and **b** the plant has grown.

4 How is energy being conserved as the plant grows?

Heat dissipation

Thermal energy always travels in one direction: it always travels from a hotter region to a cooler region.

We see **heat dissipation** in everyday life. You can feel heat dissipation as soon as you step onto cold floor tiles in bare feet, for example. The heat energy from your warm feet travels to the cooler tiles. When we cook and put a pan with cold food onto a warm cooker, both the pan and the food will become hotter.

5 Can you think of three more everyday examples of thermal energy being transferred from a hotter region to a cooler region?

There are three ways in which heat energy can travel: **conduction**, **convection** and **radiation**. Together they are known as **thermal energy transfer**.

Conduction

Materials that allow heat to pass through them easily are called **conductors** of heat. Materials that do not allow heat to pass through them easily are called **insulators**.

▲ **Figure 12.7** The conduction of heat through the bottom of a pan.

In **conduction**, heat energy is passed from one particle of a material to the next particle. When a metal pan of water is put on a stove, the atoms in the metal closest to the heat source receive heat energy first and vibrate most vigorously. Then they knock against the atoms a little further up the pan and make them vibrate more strongly too. These atoms knock against other atoms a little further up again, and the kinetic energy is passed on in this way, as shown in Figure 12.7.

Conduction can occur easily in solids but less easily in liquids, and hardly at all in gases, since the gas atoms are too far apart to affect each other. It cannot occur in a vacuum, such as outer space, where there are no particles to pass on the heat energy. Conduction is fastest in metals because they have electrons that are free to move. When a metal is heated, the electrons in that part move about faster and pass on heat energy to nearby electrons and atoms, so that the heat energy spreads quickly through to the cooler parts of the metal.

Insulators are useful in reducing the loss of heat energy. For example, a thick woollen pullover is a good insulator because the woollen fibres trap air, which is a poor conductor. This means that the pullover keeps in your body heat in cold conditions. **Insulation** materials are used in attics and lofts of houses to reduce heat loss through the roof. These materials are not only poor conductors of heat themselves, but they also trap air, which is an excellent insulator, to reduce conduction further.

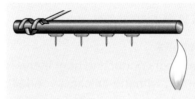

▲ **Figure 12.8**

6 A metal rod has drawing pins stuck to it with wax. It is heated at one end as shown in Figure 12.8. What do you think will happen in this experiment? Make a prediction and explain your answer.

Science extra: The structure of metals
We know that atoms come together to form covalent and ionic bonds. Atoms in metals, however, are not bonded together in this way. They lose electrons in their outer shells to become more stable and these electrons flow freely around the atoms. This is sometimes described as a 'sea of electrons' and they bind the atoms together. It is these electrons moving between the atoms that can carry electricity and heat through the metal and make it an electrical and thermal conductor.

Do all metals conduct heat at the same speed?

You will need:

rods or wires of different metals which have the same dimensions, a tripod, a Bunsen burner or other heat source, small pieces of wax, a half-metre rule, a stop-clock or timer, a cloth or tongs for handling hot items.

Work safely

Take precautions and use safety equipment when working with high temperatures.

Hypothesis

As metals are made from different atoms, they might conduct heat at different speeds.

Prediction

Examine the rods or wires and make a prediction about the order in which they will conduct heat. Begin with the rods or wires you think will conduct heat the most quickly.

Planning, investigating and recording data

Look at the equipment you will need and produce a step-by-step plan to test your prediction. Include information about what you need to do to work safely. Identify a method of recording your observations. If your teacher approves your plan, try it.

Examining the results

Compare the speeds of conducting heat in the rods or wires and arrange them in order, starting with the quickest conductor.

Conclusion

Do the results you have collected support your prediction?

LET'S TALK

Imagine you are taking part in an expedition to a cold part of the world. You have a selection of materials that can be used to make an outer coat and trousers. Work with a partner or in a group to devise an investigation to find out which material is the best insulator (poorest conductor). Share and compare your ideas with those of other groups. Do you agree on which materials you would investigate? Do you have similar or different approaches to how you would investigate the different materials? What factors are most and least important to consider in the investigation?

▲ **Figure 12.9** Polar explorers need protection from extreme cold.

Convection

In **convection**, heat energy is carried away by particles of the material that change position. When a metal pan of water is put on a stove, the water closest to the hot surface at the bottom of the pan receives heat from the metal first. The molecules of water next to the metal move faster and spread further apart as their kinetic energy increases. This makes the water at the bottom of the pan less dense than the water above it and the warm water rises. Cooler water from above moves in to take the place of the rising warmer water. The cool water is also warmed and rises. It is replaced by yet more cool water and convection currents are set up, as shown in Figure 12.10.

▲ **Figure 12.10** The convection currents in a pan of water heated from below.

Convection can only occur in liquids and gases. It cannot occur in solids where the particles are not free to move about, nor in a vacuum such as outer space.

7 When coal burns, particles of soot rise up above the fire and make smoke. Why doesn't the smoke move along the ground?

8 a The temperature of the land surface is higher than the temperature of the sea surface during the day. Use the ideas of convection currents to suggest what happens to the air above the land and above the sea.

   b Which way do you think the wind will blow across the seashore in Figure 12.11 during the day? Explain your answer.

▲ **Figure 12.11**

   c At night, the land surface is cooler than the sea surface. Does this affect the wind direction? Explain your answer.

CHALLENGE YOURSELF

Draw a diagram to show convection in a pan of boiling water. Include arrows to show the convection currents. Then write the steps to describe the process of heat transfer through the water by convection, referring to your diagram.

In a small group, share your diagram and explain convection. Listen to the explanations given by others. How are your explanations similar and how are they different?

Radiation

Thermal energy can travel through air or through a vacuum by **radiation**. For example, as a pan of water gets hotter, you can put your hand near its side and feel the heat on your skin, even though you are not touching the metal. The sides of the pan are radiating heat in the form of waves travelling through the air. These carry the heat energy from the surface of the pan to the surface of your skin, which is warmed by them.

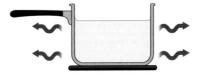

▲ **Figure 12.12** Heat radiation from a hot pan.

LET'S TALK

As a group, plan to look for examples of conduction, convection and radiation in everyday life, such as in school, on the way between school and home, at home or in any sports or other extracurricular activities that you do.

Each member of the group should bring a list of examples to the next meeting and discuss their findings. What can the group conclude about these forms of energy transfer in our daily lives?

Evaporation

▲ **Figure 12.13** Evaporation.

Particles in liquids have different amounts of energy. The particles with the most energy move the fastest. High-energy liquid particles near the surface move so fast that they can break through the surface and escape into the air and form a gas. This is called **evaporation**. When these particles leave, the amount of energy remaining in the liquid is reduced, and the temperature of the liquid drops, making it cooler. You can feel the effect of evaporation cooling your skin by moistening a finger and blowing on it.

The human body must maintain a temperature of 37 °C for good health. When we exercise, our muscles release some energy as heat and our body temperature rises. This causes sweat glands in the skin to release sweat onto the skin's surface. As the sweat evaporates, the skin's temperature falls which draws heat from the body to the surface and cools the body down.

Does evaporation cause cooling?

You will need:

two thermometers, two clamps and stands, a beaker of water, a netting-type cloth such as muslin.

Hypothesis

Look at the equipment provided and construct a testable hypothesis about how it can be used to answer the enquiry question.

Prediction

Make a prediction from your hypothesis.

Planning, investigating and recording data

Make a plan which involves a fair test with the two thermometers and a method of recording your data. If your teacher approves, try it.

Examining the results

Compare the two sets of data you have collected.

Conclusion

Do your results support your prediction?

What are the limitations of your experiment? How could you improve the experiment to make the data more reliable?

Science in context

Uses of cooling by evaporation

Cooling systems are used in fridges around the world. They bring about cooling by evaporation. Figure 12.14 shows how they work.

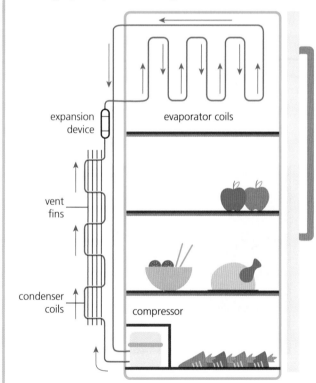

▲ **Figure 12.14** How a refrigerator works.

A fridge contains a liquid called refrigerant, which is enclosed in a system of pipes and components. As the refrigerant moves through this system, it changes from a gas to a liquid and then evaporates into a gas again. It is the evaporation of the gas which makes the fridge cool.

The refrigerant moves through the system in the following way:
- The compressor receives the refrigerant as a gas from the evaporator coils. It squashes the gas so much that its particles heat up and get closer together.
- The hot gas moves into the condenser, where the heat is removed from the gas, and it turns into a liquid. The heat passes into ventilation fins at the back of the fridge and then into the air around the fridge.
- The liquid passes into an expansion chamber, where it evaporates, and this change into a gas causes cooling. The cold gas then passes through the evaporation coils, where heat passes into it from the air and items in the fridge. This process makes the inside of the fridge and its contents cool.
- The gas then returns to the compressor and moves around the system again.

A thermostat in the fridge is a switch that is sensitive to the temperature inside the fridge and switches off the system when the temperature has been lowered enough.

9 Imagine you were a particle in the refrigerant. Describe your path around the cooling system.

Science in context

Thermal imaging

Thermal imaging is possible due to our knowledge about **infrared radiation** and its conversion into electricity to make pictures.

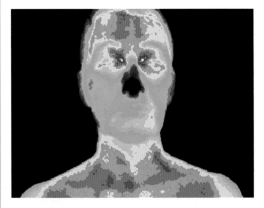

▲ **Figure 12.15** A thermographic image of a human.

Infrared radiation was discovered by William Herschel (1738–1822). In 1800, he began investigating the heat in sunlight and set up a prism to produce a spectrum with the purpose of relating light and colour to heat.

▲ **Figure 12.16** Sir William Herschel.

He tested each colour of light with a thermometer to see if the light also changed the temperature. When he tested just beyond the red light, where there was nothing to see, he found that the temperature went up the most. In Chapter 12, the transfer of thermal energy through air or a vacuum was described as radiation, and as this raising of temperature is beyond the red light, it became known as **infrared radiation**.

Over the next century, other scientists joined in the investigation of infrared radiation and discovered a way of measuring it using electricity. One of the early experimenters, Macedonio Melloni (1798–1854) made a device which could detect infrared radiation coming from a person at a distance of ten metres. Later, Samuel Pierpont Langley (1834–1906), an American scientist, invented a device that could detect radiation coming from a cow at a distance of 400 metres away.

The development of thermographic cameras followed, and today they are used around the world for seeing warm objects at night, objects hidden by smoke in a fire and in medicine to diagnose disorders in the body without having to touch people which makes these cameras patient-friendly.

▲ **Figure 12.17** Animals seen through a thermal-imaging camera.

Science extra: Why greenhouses warm up

Some infrared radiation can pass through certain solids, such as glass. For example, the infrared radiation from the Sun can pass through glass in a greenhouse but the (longer **wavelength**) infrared radiation from the ground and the plants inside the greenhouse cannot pass back out through the glass. This infrared radiation is trapped and warms the contents of the greenhouse.

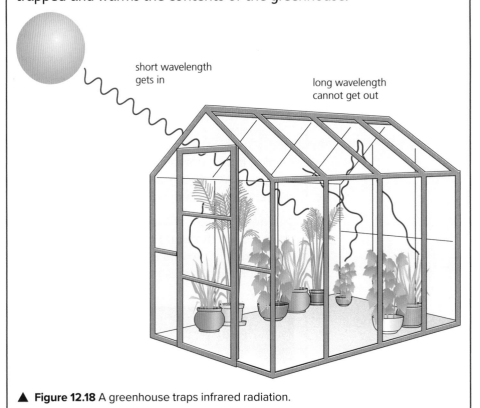

short wavelength gets in

long wavelength cannot get out

▲ **Figure 12.18** A greenhouse traps infrared radiation.

Summary

✔ Heat is a form of energy and can be transferred from one object to another.
✔ Temperature is the property of an individual object and can be measured.
✔ Energy cannot be created or destroyed, but is conserved.
✔ Heat dissipation describes the movement of heat (thermal energy) from a hotter region to a cooler region.
✔ Heat (thermal energy) can be transferred by conduction, convection and radiation.
✔ Evaporation is the process in which particles with the most energy escape from a liquid and form a gas, and this cools the liquid.
✔ Thermal imaging has been used to identify objects at night, to find objects hidden by smoke and in medicine.

End of chapter questions

1 What is another name for thermal energy?

2 What happens to the atoms and molecules in a substance when it is heated?

3 What happens to the temperature of a substance when it is heated?

4 What is the instrument normally used in a laboratory to measure temperature?

5 What does 'conservation of energy' mean?

6 What is the difference between conduction and convection in the movement of heat?

7 What happens when heat travels by radiation?

8 What happens to a moist surface when water evaporates from it? Explain your answer.

 9 a What is used to create a picture with a thermal-imaging camera?

 b State two uses of thermal imaging.

 Now you have completed Chapter 12, you may like to try the Chapter 12 online knowledge test if you are using the Boost eBook.

In this chapter you will learn:
- to use the particle model to explain how sound waves are formed when particles vibrate
- to identify the main features of a waveform
- how amplitude is linked to loudness
- how pitch is linked to frequency
- about the Doppler effect (Science extra)
- what happens when sound waves interact
- how to model sound waves
- about acoustics across the world (Science in context).

▲ **Figure 13.1** Shouting can create an echo.

Do you remember?

- How do particles move in a sound wave?
- Do sound waves travel in a vacuum? Explain your answer.
- How could you show that a sound wave is reflected?
- What is an echo?

Sound and vibrations

Look at Figure 13.2. There is a clearly great deal of sound at this live music event. The sound is being produced by vibrations. At this musical event, the vibrations take place in the amplifiers and speakers on the stage.

▲ **Figure 13.2** A live music event with a big stage and sound system.

A simple demonstration of a vibration is the 'twanging' of a ruler, as shown in Figure 13.3 on the next page.

▲ **Figure 13.3** Making a ruler vibrate.

This movement can be represented in a diagram, as shown in Figure 13.4.

▲ **Figure 13.4** Vibration is a to-and-fro movement.

The particle model and the spread of sound

We have used the particle model many times to explain how matter behaves; it can also be used to explain how sound spreads out from a vibrating object.

Think about the edge of the ruler moving up and down in the air around it. We know that the air is composed of particles that are free to move in any direction and have spaces between them. As one side of the vibrating ruler moves in one direction, towards the air particles, it pushes on them and squashes them together. The particles themselves are not compressed, but the pressure in the air at that place rises because the particles are closer together, as shown by the particles in diagram **a** in Figure 13.5 on the next page. As the side of the vibrating ruler moves back, the air particles next to it also move back, as shown by the particles in diagram **b** in Figure 13.5, then the air pressure at that place falls again.

LET'S TALK

Some people find sound difficult to understand because you cannot see it. If someone asked you, 'How do we hear a sound?', how would you explain the answer? Where would you begin? What illustration or model might you use?

1 A tuning fork has two long prongs, called tines, which vibrate quickly when they are struck. When a table-tennis ball on a thread is made to touch a vibrating prong, it swings backwards and forwards. How can this demonstration be used to explain how sound waves are made?

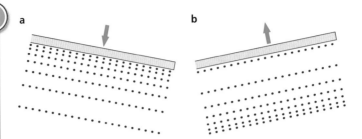

▲ **Figure 13.5** A vibrating object causes pressure variations in the air around it.

If you look at Figure 13.5 again, you can see that while the air particles are moving apart and the air pressure is falling close to the ruler, air particles further away are being squashed together and the air pressure is higher. This alternating movement of air particles to and fro, just like the side of the ruler, continues in the air further away, as Figure 13.6 shows. This movement of air particles from a vibrating object is called a sound wave.

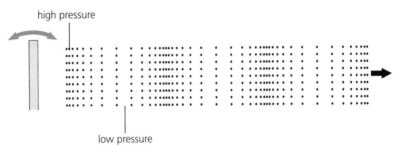

▲ **Figure 13.6** Regions of high and low pressure move away from the vibrating object.

The features of a waveform

We have seen that there are regions of high pressure and low pressure moving away from a vibrating object. If these changes in pressure are plotted on a graph, they make a waveform as shown in Figure 13.7.

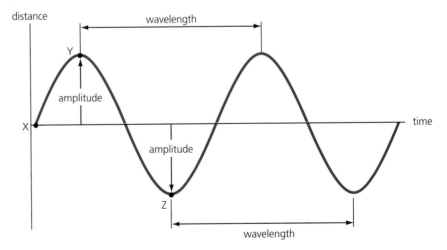

▲ **Figure 13.7** Displacement/distance waveform for a sound wave.

Figure 13.7 shows the different positions that particles can occupy when a sound wave is produced. This type of graph is called a **displacement-time graph**. A particle at position X is moving through the 'rest' position. This is the position it has when a sound wave is not being produced near it. A particle at position Y has moved the maximum distance in one direction, and a particle at position Z has moved the maximum distance in the other direction.

Two characteristics of the waveform that can be seen in Figure 13.7 are the **amplitude** and the **wavelength**. The amplitude is the maximum height of a crest (the part above the rest position) or the depth of a trough (the part below the rest position). The amplitude is measured from the rest position. The wavelength is the distance from the top of one crest to the top of the next crest, or from the bottom of one trough to the bottom of the next trough.

The loudness of a sound

The **loudness** of a sound is related to the movement of the vibrating object. If an object only moves a short distance from its rest position, it will produce sound waves with only a small amplitude, and the sound that is heard will be a quiet one. If an object moves a large distance from its rest position, it will produce sound waves with a large amplitude, and the sound that is heard will be a loud one.

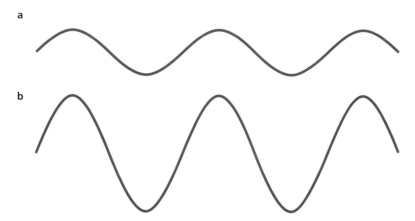

▲ **Figure 13.8** Displacement/distance waveform of **a** a quiet sound and **b** a loud sound.

The loudness of a sound is measured in **decibels**. The loudness values of different sounds are shown in Table 13.1 on the next page.

▼ **Table 13.1** The loudness of different sounds, in decibels.

Sound	Loudness/decibels
the sound hurts	140
a jet aircraft taking off	130
a road drill	110
a noisy factory floor	90
a busy street	70
normal speech	55
a whisper	20
leaves blown by wind	10
limit of normal hearing	0

▲ **Figure 13.9** Wearing ear protectors helps to prevent ear damage.

LET'S TALK

Each person in the group should listen to the sounds around them. Then using Table 13.1 to help, each person should estimate the decibel level they can hear. Everyone should compare their estimates and discuss any differences.

Loudness and energy

Sound energy passes through the air as the particles move backwards and forwards. When a wave with a small amplitude is generated, a small amount of energy passes through the air. When a wave with a large amplitude is generated, a large amount of energy passes through the air.

Sound energy can be converted into some other forms of energy. In the ear is an eardrum and three tiny bones. When a sound reaches the ear, some of its energy is changed into movement energy which moves the eardrum and bones, which then stimulates the inner part of the ear to send electrical messages through nerves to the brain.

In the microphone of a cell phone, sound energy also generates movement energy, which is converted into electrical energy, which sends information about the sound to the listener.

The pitch of a sound

You probably have an idea about the **pitch** of a sound, even if you do not know the word. You might describe a sound as being high or low, which really means a high-pitched or low-pitched. For example, when you say 'bing' you are making a higher-pitched sound than when you say 'bong'.

The pitch of the sound an object makes depends on the number of sound waves it produces in a second when it vibrates. This number of waves per second is called the **frequency**. The frequency of a sound is measured in **hertz** (*abbreviation:* Hz). The higher the frequency of the wave, the higher the pitch of the sound.

Figure 13.11 shows the positions that particles occupy at different times as the sound wave passes. Higher-frequency waves have a shorter wavelength than lower frequency waves. Sound waves share this property with waves of light.

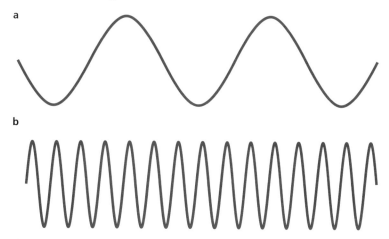

▲ **Figure 13.11** Displacement/time waveform of **a** a low-frequency sound and **b** a high-frequency sound.

The ear of a young person is sensitive to frequencies in the range of 20–20 000 hertz, but the ability to detect the higher frequencies decreases with age. Some people may have a restricted range of hearing due to nerve damage. They may not be able to hear some low-pitched or high-pitched sounds.

▲ **Figure 13.10** Cell phones use electrical energy to send the sound to the listener.

2 The following are three frequencies of sound waves: 1800 Hz, 50 Hz, 10 000 Hz.
  a Which has the highest pitch, and which has the lowest pitch?
  b What does 'Hz' stand for?

LET'S TALK

If someone asked you, 'How are loud, booming sounds made and quiet, squeaky sounds made?', how would you reply?

Is your world full of high-pitched or low-pitched sounds?

Hypothesis

Develop a hypothesis to investigate the question. It must be a hypothesis you can test.

Prediction

Make a prediction based on your hypothesis.

Planning, investigating and recording data

Make a plan to sample the sounds around you. How will you categorise the pitch of the sounds? Where will you take your sound samples and how often will you take them? How will you record your data?

Examining the results

Examine the data and make comparisons of the samples. Do they show any trends or patterns? Are there any anomalous results that you must explain?

Conclusion

Compare your analysis with your hypothesis and prediction and draw a conclusion.

What are the limitations of your conclusion? How could you improve your investigation to produce more reliable data?

3 Make drawings of the sound waves between you and the motorbike when the bike is stationary, moving at speed and then going past you.

4 Earlier, the words 'bing' and 'bong' were used to describe high- and low-pitched sounds. What words would you use to describe the different sounds of the motorbike?

Science extra: The Doppler effect

▲ **Figure 13.12** The Doppler effect causes changes to loud sounds such as motorbike engines.

Have you noticed that when a vehicle making a loud sound, such as a roaring motorbike, moves quickly past you, its sound changes? The change you hear is due to the Doppler effect. It is named after Christian Doppler (1803–1853), an Austrian mathematician.

When a motorbike is stationary with its engine running, a steady stream of sound waves is produced. As the motorbike moves at speed, the sound waves from the engine are pushed closer together, and this increases their frequency and also the pitch of the sound. As the motorbike shoots past you, the sound waves are stretched out and their frequency decreases, so the pitch of the sound falls.

When sound waves interact

We have seen that a sound wave is composed of particles under pressure when they are squashed together, and when they are pulled apart, the pressure is reduced. A region of particles under higher pressure is called a **compression**, and a region of particles under lower pressure is called a **rarefaction**.

The air has many sound waves passing through it. When two sound waves of the same frequency meet and their compressions and rarefactions occur at the same place, they are said to interfere in a **constructive** way and produce a wave of greater amplitude, as Figure 13.13 shows.

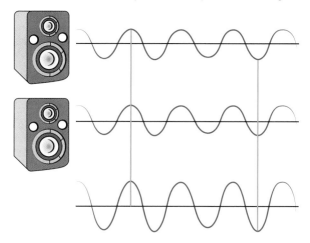

▲ **Figure 13.13** Sound waves of the same frequency interfere in a constructive way when their compressions and rarefactions occur at the same place.

When two sound waves of the same frequency meet and their compressions and rarefactions occur at different places, they are said to interfere in a **destructive** way and produce a wave with a smaller amplitude.

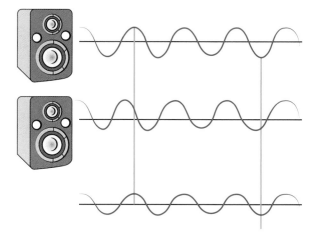

▲ **Figure 13.14** Sound waves of the same frequency interfere in a destructive way when their compressions and rarefactions occur at different places.

If two waves of the same frequency meet so that the compressions of one occur in exactly the same place as the rarefactions of the other, they cancel each other out, as Figure 13.15 shows, and no sound is heard at that place. The places produced by this total destructive wave are called **dead spots**.

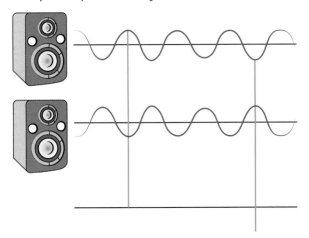

▲ **Figure 13.15** Sound waves of the same frequency can cancel each other out when the compressions of one occur in exactly the same place as the rarefactions of the other.

If two waves of different frequencies meet, they interfere both constructively (building a sound wave of greater amplitude) and also destructively (building a wave of lesser amplitude). As these two new waves move along with their different amplitudes, they produce sounds of different degrees of loudness and we hear a sound phenomenon known as **beats**.

DID YOU KNOW?

If you are tuning one musical instrument with another, you may hear beats if they are badly tuned as a waw–waw–waw sound. As you tune the instruments to make them produce sounds of the same frequency, the beats become less and eventually cannot be heard.

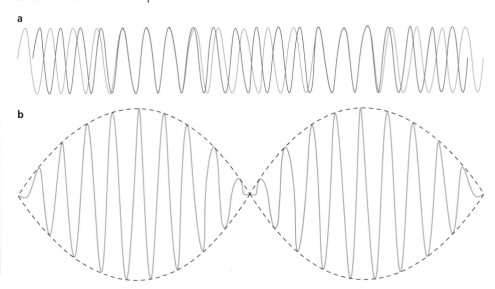

▲ **Figure 13.16 a** Two waves of different frequencies meet and **b** the resulting beats.

Modelling sound waves

Sound waves cannot be seen, but scientists consider their movement to be similar to waves moving across the surface of water. If you have ever thrown a pebble into a puddle, you will have seen waves move out across the water surface from where the pebble hit. In a similar way, the sound produced by a sound source spreads out all around it like the waves in a puddle.

Scientists use a ripple tank to model and study sound waves. An example is shown in Figure 13.17.

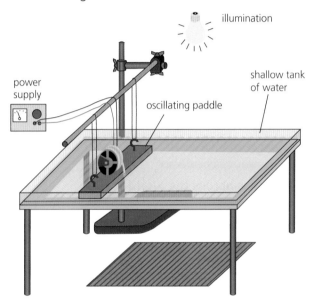

▲ **Figure 13.17** A ripple tank.

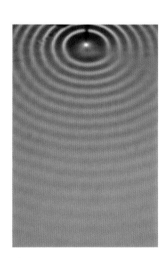

▲ **Figure 13.18** Circular wave patterns.

It is made up of a transparent tray of water to represent the puddle, and a paddle moved by a motor to represent an object generating waves. When this is switched on, ripples move across the water surface and a light shining from above casts shadows on a screen below.

When one source of movement is used, the wave pattern shown in Figure 13.18 is produced in a ripple tank. The dark areas show the positions the crests of the waves and the light areas show the positions of the troughs.

When two sources of movement are used in a ripple tank, the following wave pattern is produced. The waves from the two sources have crossed each other and interference has occurred. Where the dark regions are found are extra high crests and where light regions are found are extra low troughs. These crests and troughs produced by the interference of two waves is shown in Figure 13.19 on the next page.

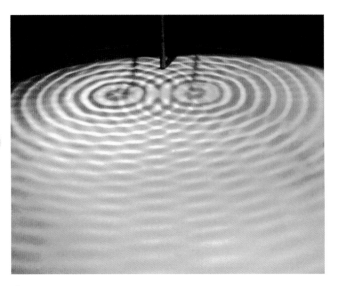

▲ **Figure 13.19** Intersecting wave patterns.

5 Does this model help you to understand sound waves? Explain your answer and think about the strengths and limitations of the model.

CHALLENGE YOURSELF

Try to make a simple ripple tank that shows how waves move across a water surface when something is dropped into it. You could use water on a tray or a plate and tap one or two spoons on it, for example. Or perhaps you could use a puddle and pebbles. Work out a plan and, if your teacher approves, try it.

Science in context

Acoustics across the world

▲ **Figure 13.20** A sound engineer at work.

Acoustics is the scientific study of sound waves in gases, solids and liquids, and applying the results of scientific enquiries in a wide variety of ways. Before studying acoustics, a person must first study physics in depth and then develop engineering skills to apply the study of sound to a particular problem. For example, when theatres and lecture rooms are being built, the acoustic engineer will look at the shape of the room and its reflecting surfaces and check that when sound waves interfere, there are no dead spots where the people on stage cannot be heard.

There are many other examples where an acoustic engineer will have tried to control sound to improve our everyday lives. A few examples are:
- studying sounds made by aircraft to help reduce noise pollution
- studying the echoes produced on voice calls and video chats on the internet to remove them
- working on the designs of headphones and loudspeakers to improve the way they deliver sound.

LET'S TALK

Imagine you have been asked by a phone company to create a new design for some headphones, to be sold with the latest model of mobile phone. What would your headphones look like? Would they be wired or wireless and why? Would they be in-ear or over-ear design, why? How would they connect to a device and how would they be controlled by the user? What shape would they be, why? Have a go at designing some headphones. Then in small groups share, compare and discuss your designs.

▲ **Figure 13.21** Several different designs for headphones.

Summary

- ✔ Sounds are caused when something vibrates.
- ✔ The particle model can be used to explain how sound travels through matter such as air.
- ✔ Two features of a waveform are its amplitude and its wavelength.
- ✔ The amplitude of a waveform is linked to the loudness of the sound made.
- ✔ The wavelength of a waveform is linked to the pitch of the sound made.
- ✔ When sound waves meet, they can reinforce or cancel out each other.
- ✔ The way that sound waves reinforce or cancel out each other can be visualised by using a ripple tank.
- ✔ Knowledge of acoustics has been applied in controlling noise pollution, studying echoes and in electronic communication.

End of chapter questions

1 What happens to air particles when an object vibrates?

2 Describe the amplitude and wavelength of a sound that
   a is loud and high-pitched
   b is loud and low-pitched
   c is quiet and high-pitched.

3 When the amplitude of a sound wave is increased, what happens to the sound it produces?

4 When the wavelength of a sound wave is shortened, what happens to the sound it produces?

5 a When can sound waves reinforce each other or cancel each other out?
   b What happens when sound waves
      i reinforce each other?
      ii cancel each other out?

 Now you have completed Chapter 13, you may like to try the Chapter 13 online knowledge test if you are using the Boost eBook.

Electrical circuits

In this chapter you will learn:

- about the differences between simple circuits and parallel circuits
- to measure current and voltage in simple circuits and parallel circuits
- what happens to current and voltage when cells and lamps are added to series and parallel circuits
- to calculate resistance
- to draw circuit diagrams that include fixed and variable resistors and buzzers
- to make and compare circuits that contain fixed and variable resistors and buzzers
- about superconductors (Science in context).

Do you remember?

- Name a good conductor of electricity.
- Name an electrical insulator.
- If you make a circuit and it does not conduct electricity, what do you need to check?
- What is used to measure the current in a circuit?
- What is the name of the tiny parts of an atom that make a current of electricity?

DID YOU KNOW?

Electricity travels at the speed of light – about 300 000 kilometers per second.

a b

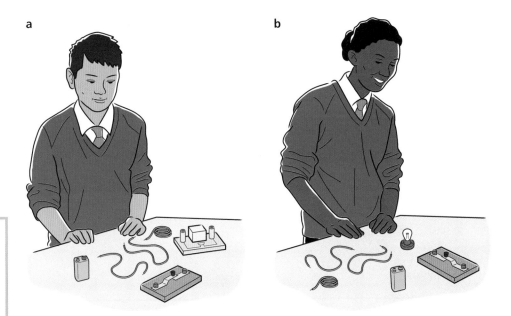

▲ **Figure 14.1** Electrical components for a circuit, featuring **a** a buzzer and **b** a lamp.

Circuits

Electrical circuits can be very complicated, as shown by Figures 14.2, 14.3 and 14.4.

It does not matter how complicated the electrical circuit is; they all work due to the properties and processes of simple circuits that we use in school science laboratories. This study of simple electrical circuits can lead some students to have an interest in a career in electrical engineering or in developing circuits like those shown.

▲ **Figure 14.2** A complicated wiring circuit.

▲ **Figure 14.3** Electronic microcircuits.

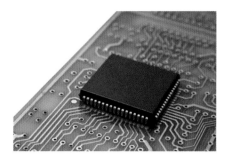

▲ **Figure 14.4 Microchip** circuits.

A simple electrical circuit

Figure 14.5 shows a simple circuit you may already be familiar with from prior learning. Attempt the questions about it to see what knowledge of electrical circuits you already have.

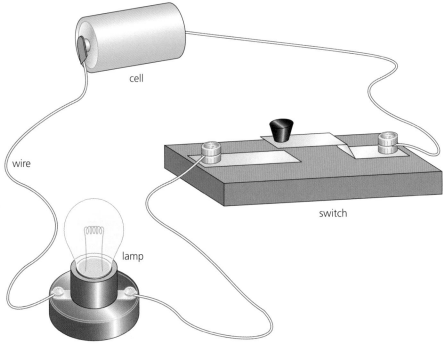

1 How do you
   a close this circuit
   b open this circuit?
2 What happens when the circuit is
   a closed
   b opened?

▲ **Figure 14.5** A simple circuit.

3 Use the symbols in Figure 14.6 to make a circuit diagram of the simple circuit in Figure 14.5.

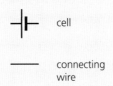 cell

connecting wire

lamp

switch

▲ **Figure 14.6** Common circuit diagram symbols.

When two or more cells are joined together, they make a battery. The symbols in Figure 14.7 show the arrangement for batteries with two cells, three cells and any number of cells.

a   b   c

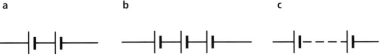

▲ **Figure 14.7** The circuit symbols for two cells, three cells and any number of cells.

Series and parallel circuits

There are two kinds of electrical circuits: **series circuits** and **parallel circuits**.

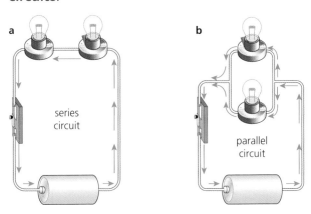

▲ **Figure 14.8 a** A series and **b** a parallel circuit

In a series circuit, all the components are arranged in a loop, as Figure 14.8 a shows. In a parallel circuit, two or more components are wired up side by side, as Figure 14.8 b shows.

Measuring current

When the **current** flows through a series circuit, you can think of it as passing through all the components one after the other. When the current flows through a parallel circuit, it divides where the components are in parallel. The two currents flowing through each part of the parallel circuit are equal to the total current that flows from the cell or battery.

The unit for measuring the current flowing through a circuit is the **ampere**. It is usually shortened to the word amp or amps and its symbol is A.

4 Use the symbols in Figure 14.6 to make circuit diagrams of Figures 14.8 a and b.

5 You have a circuit with one cell and two lamps in series.
   a How does the brightness of the lamps change if you add a third in series? Explain your answer.
   b How does the brightness of the lamps change when you add a second cell in series? Explain your answer.

We have seen that the source of the electric current in a circuit is the cell. If a circuit has one cell and a second cell is added, the current flowing through the circuit increases. The brightness of a lamp depends on the size of current flowing through it. A large current makes the lamp shine brightly and a small current makes the lamp shine with a dim light. A circuit with one lamp may shine brightly but when a second lamp is added in series, both lamps shine less brightly as the current flowing through them is reduced.

The current flow in the circuit is measured using an **ammeter**. When it is placed in a circuit, its red terminal must be connected by a wire to the positive terminal of the cell, battery or power supply. It is always connected in series with the component through which the current flow is to be measured, as Figure 14.9 shows.

LET'S TALK

Can you remember how many electrons flow past any point in a circuit in a second when the current is one amp? Is it
a   six million
b   six million million
c   six million million million
d   six million million million million?

Work in a group to discuss your ideas and agree on an answer.

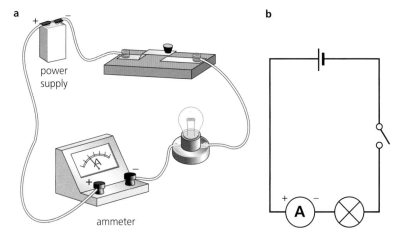

▲ **Figure 14.9 a** An ammeter connected in a circuit and **b** the circuit diagram showing its symbol.

Sometimes scientists make investigations to check what they have read. For this enquiry, set up the circuits shown in Figure 14.10 on the next page, but include a switch to find out about the current flow in them.

Can you investigate and measure current in series and parallel circuits?

You will need:

a cell, two lamps, four wires, a switch, an ammeter.

Planning, investigating and recording data

Set up each circuit in turn and measure the current at the points shown.

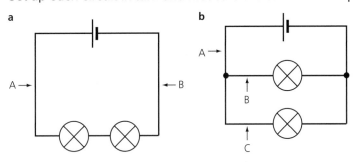

▲ **Figure 14.10** Measuring current in **a** series and **b** parallel circuits.

Record your measurements. Think about how many measurements you need to make and how you will record them.

Examining the results

Compare the measurements you made in both circuits.

Look back at the information about current in parallel circuits. How do your measurements compare to the text?

Conclusion

Draw a conclusion from your evaluation and state its limitations. For example, have you made enough measurements? Suggest how the investigation could be improved to provide more reliable data.

CHALLENGE YOURSELF

How would you measure the current produced by a lemon battery? Work out a plan and, if your teacher approves, try it.

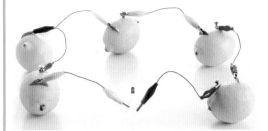

▲ **Figure 14.11** A lemon battery.

Measuring voltage

The ability of the cell to drive a current is measured by its **voltage**. This is indicated by a figure on the side of the cell with the letter 'V' after it. The volt, symbol V, is the unit used to measure the difference in electrostatic potential energy (usually just referred to as potential difference) between two points. The voltage written on the side of the cell refers to the difference in potential between its positive and negative terminals. It is a measure of the electrical energy that the cell can give to the electrons in a circuit.

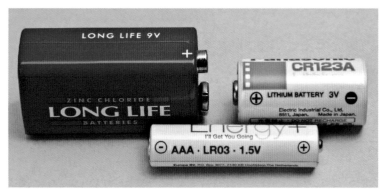

▲ **Figure 14.12** The voltage is clearly displayed on the packaging of cells and batteries.

When cells are arranged in series, with the positive terminal of one cell connected to the negative terminal of the next cell, the current-driving ability of the combined battery of cells can be calculated by adding their voltages. For example, two 1.5 V cells in series produce a voltage of 3 V. The two cells together give the electrons in the circuit twice as much electrical energy as each one would provide separately.

The voltage or potential difference between two points in a circuit is measured using a **voltmeter**.

6 Examine how an ammeter and a voltmeter are connected into a circuit by comparing Figures 14.9 and 14.13. How are they similar and how are they different?

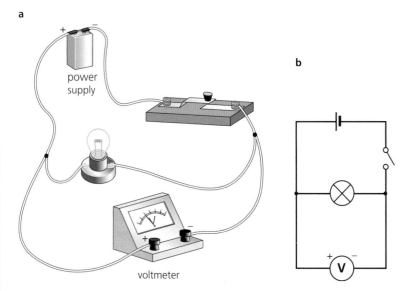

▲ **Figure 14.13 a** A voltmeter connected in a circuit and **b** the circuit diagram showing its symbol.

The voltmeter is connected into a circuit with its positive (red) terminal connected to a wire that leads towards the positive terminal of the cell, battery, or power pack. The negative (black) terminal must be connected to a wire that leads towards the negative terminal of the source of the current. However, unlike the connection of an ammeter, the wires are attached to either side of the part of the circuit being tested – it is arranged in parallel with this part of the circuit.

Can you measure and compare voltages?

You will need:

two cells, a switch, two lamps, nine wires, a voltmeter.

Here are some science enquiries related to circuit diagrams. Read the instructions and, before you begin, think about how you will record your measurements.

1   Draw a circuit diagram to show how you would check the voltage of a cell in a circuit with a switch and a lamp. Check it with your teacher and, if approved, try it.
    Look at the voltage of the cell you are to test. When you make your test, compare your voltage reading with the voltage stated on the cell. What do you find?

2   Draw a circuit diagram with a second lamp arranged in series in your previous circuit. Show how you would check the voltage of the two lamps and the cell in the circuit and, if your teacher approves, try it. What do you find?

3   Draw a circuit diagram with two lamps and two cells arranged in series in your previous circuit. Show how you would check the voltage of the two lamps and the two cells in the circuit and, if your teacher approves, try it. What do you find?

4   Draw a circuit diagram with a cell and two lamps arranged in parallel. Show how you would check the voltage of the two lamps and the cell in the circuit and, if your teacher approves, try it. What do you find?

Conclusion

Can you summarise the effect on the voltage of adding cells and lamps from the measurements you have recorded in these scientific enquiries?

Resistance

From prior learning and studies of lamps with circuits, you may know that electrons push their way through wires and components in circuits. The wires and components of circuits are made from materials which slow down the electrons and make them rub and push against the material as they move along. This property of a material to oppose the movement of electrons is called **resistance**. You can model this action by rubbing your hands together and generating heat, just as it occurs in materials with a high resistance. Try to remember your knowledge of resistance with lamps as you conduct the following scientific enquiry.

Can you investigate resistance using lamps?

You will need:

a cell, wires, a switch, three lamps.

Hypothesis

Construct a testable hypothesis to explain what might happen when a circuit is set up with one lamp, then another lamp is added in series and then a third lamp is added in series.

Prediction

Make a prediction based on your hypothesis.

Process

1   Make circuit diagrams of
    a   a cell, a switch and one lamp
    b   a cell, a switch and two lamps
    c   a cell, a switch and three lamps.
2   Set up each circuit in turn, close the switch and record the result by giving a short description of the light coming from the lamps, such as 'dim', 'bright', 'very bright'.

Examining the results

Compare the descriptions of the brightness of the lamps in the different circuits you have made.

Conclusion

Compare your results with your hypothesis and prediction and draw a conclusion.

Is your conclusion limited in some way? Explain your answer.

What improvements could be made? Explain the changes that you suggest.

Calculating resistance

In the previous scientific enquiry, resistance was observed using the light from lamps, but scientists have found a better way of measuring resistance. They have found that it can be linked to the voltage and current in the circuit in this equation:

$$\text{resistance} = \frac{\text{voltage}}{\text{current}}$$

We have seen that the unit for measuring current is the ampere or amp (it has the symbol I) and the unit for measuring the potential difference between two points in a circuit is the volt (its symbol is V). The unit for measuring resistance is the **ohm**. It has the symbol **R**. We can represent the equation in symbols as:

 $R = \dfrac{V}{I}$

The resistance in ohms of a lamp can be found by setting up a circuit, then measuring the current with an ammeter and the voltage with a voltmeter. These values can then be put into the equation to find the resistance. The ammeter reads the current running through the **resistor** while the voltmeter reads the voltage across it. A plot of voltage against current gives a measurement of the resistance.

7 A circuit is set up to find the resistance of a lamp. The voltage is found to be 3.0 V and the current is found to be 0.4 A. What is the resistance of the lamp?

▲ **Figure 14.14 a** An ammeter and **b** a voltmeter.

Each component in a circuit has a particular resistance. It may not be the same as other components. This means that the current flowing through it will be different to the current flowing through other components. The current flowing through a component can be found by rearranging the equations:

 $\text{current} = \dfrac{\text{voltage}}{\text{resistance}} \rightarrow I = \dfrac{V}{R}$

Using this equation, by keeping the voltage in the circuit the same and measuring the resistance in components, we can calculate the flow of current through the component.

For example:
- Component A has a resistance of 6 Ω and the voltage of the circuit is 12 V. Using the equation, the current = 12 V ÷ 6 Ω = 2 A.
- Component B has a resistance of 3 Ω and the voltage of the circuit is still 12 V. Using the equation, the current = 12 V ÷ 3 Ω = 4 A.

CHALLENGE YOURSELF

Draw a circuit diagram to measure the resistance of a lamp. If your teacher approves, set up your circuit and work out the resistance of the lamp.

From these examples it can be seen that the higher the resistance of the component, the smaller the current flowing through it.

8 a   What is the current flowing through component X of $4\,\Omega$ resistance in a circuit at $12\,V$?

b   When component Y of $2\,\Omega$ resistance is placed in the circuit at $12\,V$, does it have more or less current flowing through it than component X? Explain your answer.

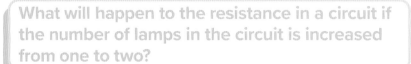

What will happen to the resistance in a circuit if the number of lamps in the circuit is increased from one to two?

You will need:

a selection of electrical equipment.

Hypothesis

Construct a testable hypothesis to explain what will happen to the resistance in a circuit if the number of lamps in the circuit is increased from one to two.

Prediction

Make a prediction based on your hypothesis.

Process

1   Construct a circuit diagram to investigate the resistance of a circuit with one lamp in it.

2   Construct a circuit based on your diagram and, if your teacher approves, measure the voltage and current and use the equation to calculate the resistance.

3   Construct a circuit diagram to investigate the resistance of a circuit with two lamps in it.

4   Construct a circuit based on your diagram and, if your teacher approves, measure the voltage and current and use the equation to calculate the resistance.

Examining the results

Compare the results of your calculations.

Conclusion

Compare your results with your hypothesis and prediction and draw a conclusion.

What are the limitations of your conclusion? How could you improve your investigation to produce more reliable data?

LET'S TALK

What have you found out about electricity from your studies on circuits up to this point in the chapter?

In a group, spend a little time going through the previous pages, then make a careful step-by-step presentation of your ideas. Give your presentation to the class and allow others to give feedback when you have finished.

Fixed resistors

The wire in lamps becomes so hot that it gives out light. The heat makes the wire change its resistance as it gets hotter. In electronic circuits, components called **resistors** are used. They maintain their resistance and are not affected by heat. They are also called **fixed resistors** and are used in further studies on resistance.

A component that is designed to introduce a particular resistance into a circuit is called a resistor.

▲ **Figure 14.15 a** Four resistors and **b** the symbol for a fixed resistor.

Can you measure the current and voltage and calculate resistance?

You will need:

a fixed resistor and a selection of electrical equipment.

Hypothesis

Resistance can be found by using the equation:

$$resistance = \frac{voltage}{current}$$

Prediction

Measuring the voltage and current will allow the resistance in the circuit to be calculated.

Planning, investigating and recording data
1 Draw a circuit diagram. This time, instead of including a lamp, put a resistor in the circuit.
2 Use your diagram to select the equipment you will need.
3 Assemble the circuit using your diagram to help you.
4 Switch on your circuit and record your measurements.

Examining the results

Use your measurements of current and voltage to calculate the resistance.

Conclusion

Compare the examination of your result with the hypothesis and prediction.

Is your conclusion limited in some way? Explain your answer.

What improvements could be made? Explain the changes that you suggest.

Variable resistors

A **variable resistor** can be made in which a contact moves along the surface of a resistance wire and brings different lengths of the wire into the circuit. In order to make it more compact, the length of wire is wound in a coil and the contact is made to move freely across the top of the coil.

In Figure 14.16, the current passes through terminal A, along the bar, through the sliding contact and coil of wire to terminal B. When the contact is placed on the far left, the current passes through only a few coils of the wire. As the contact is moved to the right, the current flows through more of the wire and encounters more resistance. When the contact is moved from the right to the left, the current flows through fewer coils of the wire and encounters less resistance.

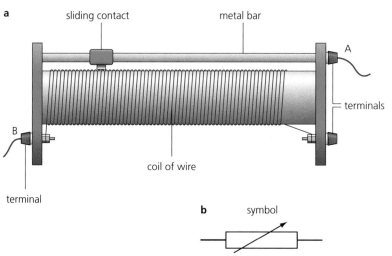

a

sliding contact   metal bar

A

terminals

B

coil of wire

terminal

b   symbol

▲ **Figure 14.16 a** A variable resistor and **b** the symbol for a variable resistor.

9 In Figure 14.16, which way should the contact be moved to
   a increase
   b decrease
   the resistance in the part of the wire included in the circuit between A and B?

10 Figure 14.17 shows a variable resistor in a dimmer switch. How would you turn the switch to make the lights
   a brighter
   b dimmer?

   Explain your answers.

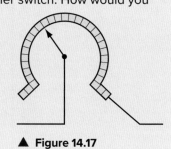

▲ **Figure 14.17**

Can you measure the resistance in a circuit produced by a variable resistor?

You will need:

two cells, an ammeter, a lamp, a variable resistor, a switch.

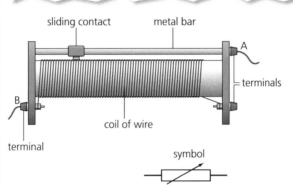

▲ **Figure 14.18** A variable resistor set up in a simple circuit.

Hypothesis

Construct a testable hypothesis to explain what you think will happen when you set up a circuit using the equipment listed above.

Prediction

Construct a prediction based on your conclusion.

Planning, investigating and recording data

Make a circuit using the equipment listed above.

Make a plan of how you will carry out the investigation and, if your teacher approves, try it. Be sure to record data as you investigate.

Examining the results

Compare different sections of your data.

Conclusion

Compare your evaluation with your hypothesis and prediction. What do you conclude?

What are the limitations of your conclusion? How could you improve the investigation?

Buzzers

A **buzzer** is an electrical device with a part that vibrates strongly when a current of electricity passes through it. The vibrations produce a sound.

▲ **Figure 14.19 a** A buzzer and **b** the symbol for a buzzer.

What might happen when you use a variable resistor in a circuit with a buzzer?

You will need:

a selection of electrical components to investigate the effect of a variable resistor in a circuit, a decibel meter app on a cell phone (optional).

Hypothesis
Construct a testable hypothesis to answer the question.

Prediction
Make a prediction based on your hypothesis.

Planning, investigating and recording data
1 Construct a circuit diagram you wish to use in your investigation.
2 Write down the steps you will take in your investigation. State what you will record and how it will be presented.
3 Select the electrical components you will need.
4 Show your teacher your diagram plan and selected components and, if approved, make your investigation.

Examining the results
Do your results show a trend or pattern? Are there any anomalous results?

Conclusion
Compare your evaluation with your hypothesis and prediction and draw a conclusion.

What are the limitations of your conclusion? How could the investigation be improved?

Science in context

Superconductors
Heike Kamerlingh-Onnes (1853–1926) was a Dutch physicist who studied how gases behaved as they turned into liquids. He was particularly interested in helium, which has a boiling point of −269°C. In 1908, he cooled helium gas so much that it turned into a liquid.

Kamerlingh-Onnes set up a cryogenic laboratory at the University of Leiden and began studying how other materials behaved at low temperatures. In 1911, he cooled mercury to a temperature of almost −273°C and passed electricity through it. He found that the metal offered

▲ **Figure 14.20** The magnetic field around the superconducting plate (the white block) makes the magnetic disc (the small floating object) float in the air.

no resistance to current. The property of a material that offers no resistance to current is called **superconductivity** and the material is called a **superconductor**. Other metals that behave as superconductors at very low temperatures include aluminium and lead.

When a current is passed through a normal conductor, energy is needed to overcome the conductor's resistance. When a current is passed through a superconductor, no energy is needed to overcome resistance because the resistance does not exist.

Since Kamerlingh-Onnes' time, scientists have studied many materials and tested them for superconductivity. They have found some materials that behave as superconductors at higher temperatures. In 1987, Alex Müller (born 1927), a Swiss physicist, and J. Georg Bednorz (born 1950), a West German physicist, shared the Nobel Prize for Physics for discovering that a ceramic material containing barium and copper oxide behaves as a superconductor at −238 °C.

Since then, more materials have been discovered that are superconductors at higher temperatures, but until superconductors can be used at normal temperatures, their use will remain limited. Today, they are used in particle accelerators, which are used to research the particles in atoms, and certain types of body scanners used in hospitals.

LET'S TALK

Suggest how you think the use of superconductors at normal temperatures could affect the use of energy resources. Share your ideas with a partner.

11 What was Kamerlingh-Onnes' first interest in scientific investigations?

12 Why did his interest in helium lead him to study materials at low temperatures?

13 Compare the behaviour of a superconductor with that of an ordinary conductor.

14 What extra equipment is needed if a superconductor is to be used in a circuit?

Summary

✔ There are two kinds of circuit: series and parallel circuits.
✔ An ammeter is used to measure the current in both series and parallel circuits.
✔ The voltage is the difference in potential between two points in a circuit and is also the ability of a cell to drive a current around a circuit.

✔ A voltmeter is used to measure voltage in both series and parallel circuits.
✔ Voltage and current in a circuit change when cells and lamps are added.
✔ Resistance is the property of a material to oppose the movement of electrons.
✔ Resistance can be calculated with the use of the equation:

$$\text{resistance} = \frac{\text{voltage}}{\text{current}}$$

✔ Fixed resistors and variable resistors are used to control the flow of electric current in a circuit.
✔ A buzzer is a component that makes a sound when electricity passes through it.
✔ Superconductors can be used to conduct a current with little or no resistance, but they need very low temperatures and are mainly used to investigate the structure of atoms and in some hospital body scanners.

End of chapter questions

1 Make a list of the components in each of the three circuits shown in Figure 14.21.

a 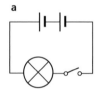 b 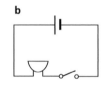 c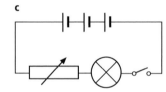

▲ **Figure 14.21**

2 Draw the following circuit diagrams:
   a a simple series circuit with two lamps in it
   b a simple parallel circuit with two lamps in it.

3 Describe the path of the current in
   a the series circuit you have drawn in your answer to Question 2a
   b the parallel circuit you have drawn in your answer to Question 2b.

4 What is an ammeter used for and how do you set it up in a circuit?

5 What does the word 'voltage' mean?

6 When you use a voltmeter in a circuit, do you set it up in exactly the same way as an ammeter? Explain your answer.

7 What do scientists mean when they talk about the 'resistance' of a wire?

8 a If a current of 2 amps flows through a 60-volt lamp, what is the resistance of the lamp?
   b If a current of 4 amps flows through a 60-volt lamp, what is the resistance of the lamp?

Now you have completed Chapter 14, you may like to try the Chapter 14 online knowledge test if you are using the Boost eBook.

Planet Earth

In this chapter you will learn:

- how ideas about the Earth have changed over time (Science in context)
- how evidence for tectonic plates comes from rocks, the ocean floor, fossils, the position of volcanoes and earthquakes, and the alignment of magnetic materials
- about cratons (Science extra)
- how tectonic plates move.

Do you remember?

- Rocks can be classified in three ways. What are they?
- What is the outer layer of the Earth called that forms its surface?
- There is another layer beneath the outer layer. What is that called?
- Are you sitting on a tectonic plate? Explain your answer.
- Where are most earthquakes, volcanoes and fold mountains found?
- Does Earth have a magnetic field? Explain your answer.
- What do both Figures 15.1 a and b show?

▲ **Figure 15.1** What can you see in these images?

Science in context

Changing ideas about the Earth

When explorers first began to travel the world, they made maps of the coastlines of lands they discovered. In 1596, Abraham Ortelius (1527–98) constructed a world map and noticed that large areas of land (the continents) around the Atlantic Ocean looked like they could have fitted together like a jigsaw at some earlier time. He believed that the land had been pushed or pulled apart by floods and earthquakes.

▲ **Figure 15.2** An early map of the Earth.

1 How does the early map of the Earth compare with the modern map of the Earth? Give five examples of the differences.

In 1666, Francois Paget believed that the large areas of land may have become separated because land between them had sunk into the ocean. In 1858, Antonio Snider-Pellegrini (1802–85) believed that at one time all the large areas of land were joined together to make a supercontinent.

▲ **Figure 15.3** A modern map of the Earth – you can see the shapes of the continents.

Evidence for tectonic plates

Over time, scientists have used a range of evidence to investigate the idea that there was once a supercontinent and that land has split and moved to form the continents that we have today. But where does this evidence come from?

Evidence from rocks

LET'S TALK
How would you explain to someone that all rocks are not the same and there are three different types? Work in groups to discuss your ideas.

Scientists have classified rocks into three groups: **igneous**, **sedimentary** and **metamorphic** rocks.

Alfred Wegener (1880–1930), a geophysicist from Germany, studied rocks and fossils from around the world. He discovered similarities in some of them which led to increased development of our knowledge and understanding of the continents.

Igneous rocks are formed from molten rock called **magma**, which comes from beneath the Earth's crust. After magma rises to the surface, it cools down to form igneous rock.

Sedimentary rocks are mostly formed from fragments of rock which are produced by **weathering** and **erosion**. They can be transported by water or wind and can be packed together over time. Sandstone is an example of a sedimentary rock.

▲ **Figure 15.4** Sandstone is formed when tiny grains of rock are pressed and bound together.

Many sedimentary rocks are made from the shells of ancient living things. For example, limestone is formed from sea creatures, such as molluscs, which fell to the ocean floor when they died and their shells packed together.

2   What do the words 'weathering' and 'erosion' mean?

Some sedimentary rocks are formed from dried up seas. Seawater contains many chemicals which have dissolved out of the minerals in the rocks and been washed down rivers into seas and oceans. In the past, many seas have dried up and in that process water evaporated from the surface, leaving the minerals behind to make the remaining seawater more concentrated. Eventually, there was not enough water left for all the chemicals to remain in the solution, and some of them joined together to form crystals. Rocks that form this way are called **evaporites**, and rock salt is an example.

▲ **Figure 15.5** You can see the shells in this piece of limestone.

Metamorphic rocks are formed from igneous and sedimentary rocks that have been heated and squashed in the Earth's crust. A rise in temperature and pressure causes rocks to change their form (or metamorphose). For example, when limestone is heated and squashed in the Earth, it turns into marble.

▲ **Figure 15.6** Rock salt is also called halite.

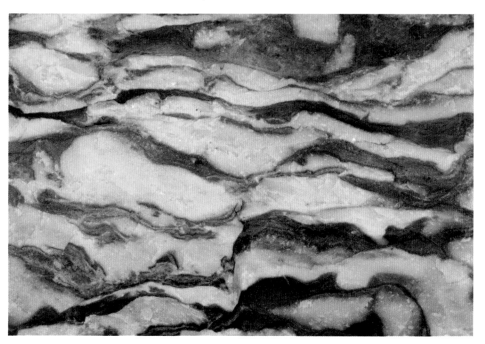

▲ **Figure 15.7** Marble.

LET'S TALK

If you found two samples of rock that were exactly the same but on two different continents, what conclusions could you draw? Work in a group. Discuss your ideas and then share your ideas with the class.

Science extra: Cratons

A **craton** (sometimes also called a shield) is an area of rock which has not been disturbed by movements of the Earth's surface for between half a billion and three and a half billion years. They originally formed from igneous or metamorphic rock which was then covered with sedimentary rock.

On the west coast of Africa and the eastern coast of South America, cratons of the same type of rock have been found that formed at the same time over two billion years ago. This suggests that at that time they were on the same land-mass, and they could only be found so far apart now as a result of the land spitting into pieces and moving away from each other.

Evidence from under the oceans

Investigations of the floor of the Atlantic Ocean have shown that it has a mountain ridge running down the middle of it from pole to pole. Figure 15.8 is a bathymetric map of the Atlantic Ocean (a special kind of map that shows the underwater features of oceans and seas). You can clearly see the Mid-Atlantic Ridge running through the middle. The Mid-Atlantic Ridge is an example of a boundary where the **tectonic plates** have moved apart.

In some places, there is a large, distinct gap down the middle of the ridge. Divers can swim between the two plates in such places.

▲ **Figure 15.8** The Mid-Atlantic Ridge.

▲ **Figure 15.9** A diver swimming between two tectonic plates.

Evidence from fossils

Fossils are formed of the dead bodies of plants and animals that have been quickly covered after death by sand or mud. This covering prevents them from being eaten by animals feeding as scavengers, and the lack of oxygen around them stops **decomposers** from breaking down their bodies. In time, water and minerals passing through the rock turn their bodies into fossils. Movements of Earth's crust can expose the fossils and this makes them easier for scientists to find. In some places, fossils can be seen in rocks along coastlines, such as in Lyme Regis, which is found on the Jurassic Coast in the south of England.

▲ **Figure 15.10** The result of fossilisation.

The study of fossils from different continents provides evidence that the continents have moved apart. Here are some examples:

● Fossils of a freshwater, crocodile-like reptile called mesosaurus, that lived between 286 and 258 million years ago, have been found in South America and Africa. As the reptile could not live in seawater and the chance of two identical species evolving at the same time in two different places is extremely unlikely, the explanation for their distribution is that they all lived on the same land which later split apart.

▲ **Figure 15.11** Mesosaurus fossil.

- Fossils of an animal with characteristics of reptiles and mammals, called cynognathus, lived between 250 and 240 million years ago and has been found in South America and South Africa.

▲ **Figure 15.12** Reconstruction from cynognathus fossils.

- Fossils of a woody plant, called glossopteris, which lived between 299 and 250 million years ago have been found in South America, South Africa, Australia and Antarctica.

LET'S TALK

Why and how do fossils provide evidence of tectonic plates? Work in a group and discuss your ideas. Hint: begin by describing what a fossil is and build on your answer from there.

▲ **Figure 15.13** Reconstruction from glossopteris fossils.

 ## Evidence from volcanoes and earthquakes

Volcanoes and earthquakes occur in specific locations on Earth. These locations are related to tectonic plates, and specifically to the places where two or more tectonic plates meet, called plate boundaries. At these boundaries, earthquakes and volcanoes will occur as a result of the movement of the tectonic plates. Plates can be moving apart, moving towards each other, or one plate may be submerged (subducted) below another. Plates can also rub alongside each other which is why many earthquakes occur in San Francisco, California. The city is located where the Pacific Plate and the North American Plate rub against each other causing friction to build. When the friction is released, the plates jerk and we feel this on the surface as an earthquake.

Volcanic eruptions happen when molten rock called magma escapes from below the crust onto the surface or into the sea. After magma is released onto the surface (or into the sea) during a volcanic eruption, it is called lava. Some volcanic eruptions are huge, explosive forces and some are more gentle, regular lava flows. Magma is also able to reach the surface through gaps or openings at the boundary between tectonic plates. Here, lava can cool and solidify to form new land, for example at the Mid-Atlantic Ridge, where the North American and Eurasian Plates move apart. This continuous movement and constant cooling and solidifying causes the land mass of Iceland to grow by approximately 2.5 cm every year.

3 Look at Figure 15.14. Notice how groups of volcanoes can often be found in lines or chains. Why do you think this is the case?

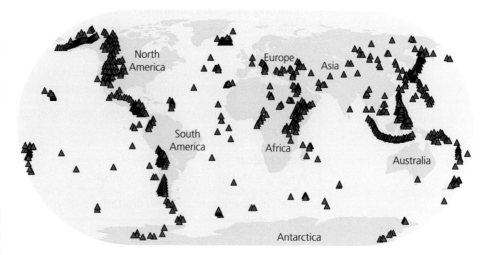

▲ **Figure 15.14** Worldwide volcano distribution.

LET'S TALK

Use the internet to find a map of the major tectonic plates of the Earth. Compare it to the world map in Figure 15.14. What do you notice? Share your ideas with a partner or in a group.

Evidence from magnetism

When lava cools and solidifies, the Earth's magnetism causes minerals in the rock to line up, pointing to one of the poles. This occurs in the rock on both sides of the boundary and is evidence that they were produced at the same time. This is called magnetic polarity. The Earth's magnetism is not constant; when it changes to the opposite pole, the minerals in the rock point in the opposite direction. This is called reverse magnetic polarity. This repeats and produces a pattern of magnetic stripes, as you can see in Figure 15.15. Geologists can use this pattern to calculate the rate at which new crust is being made.

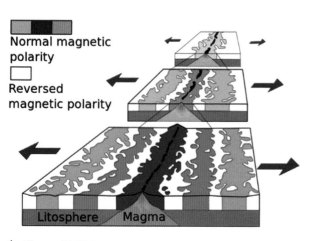

Normal magnetic polarity

Reversed magnetic polarity

Litosphere   Magma

▲ **Figure 15.15** A pattern of magnetic stripes around the Mid-Atlantic Ridge.

4 How can magnetism help us to understand the speed of movement of the tectonic plates?

DID YOU KNOW?

Scientists who study the structure of the Earth and the rocks the planet is made from are called geologists.

How can you model a supercontinent jigsaw?

You will need:

some paper or cardboard, scissors, modelling clay, rock specimens (optional).

Planning and investigating

Make a model to show how evidence from rocks and fossils demonstrates how separate continents were once part of a supercontinent.

You do not need to use the precise shapes of the Earth's continents; you can make a model with simple geometric shapes instead.

When your model is complete, demonstrate its use and perhaps film your demonstration.

Conclusion

What are the analogies in your model?

What are the strengths and limitations of your model?

How do tectonic plates move?

We have seen how the movement of tectonic plates can result in volcanoes and earthquakes, but what causes the plates to move?

Tectonic plates are the outermost layer of planet Earth, they form the Earth's crust. Below the crust is a layer called the mantle, which is filled with magma. Magma is molten rock, which takes the form of a thick, viscous liquid. The two innermost layers make up the Earth's core.

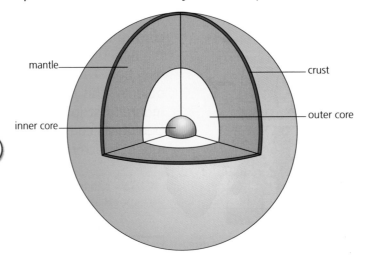

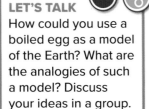

LET'S TALK
How could you use a boiled egg as a model of the Earth? What are the analogies of such a model? Discuss your ideas in a group.

▲ **Figure 15.16** The major components of the structure of the Earth.

CHALLENGE YOURSELF

Boundaries between plates are given different names according to the way in which they move. Do some research to find about about the names of different types of boundaries and the movement taking place there. Write some notes about what you discover and then share your findings in a group. You may find it useful to sketch diagrams to show how the plates are moving.

Temperatures in the mantle are not consistent; it is hotter in areas closer to the core than areas closer to the crust. Close to the core, temperatures can reach almost 4000 degrees Celsius. Areas below the crust reach up to approximately 1000 degrees Celsius. As a result of the variations in temperature, thermal convection occurs. This is where heat rises within the mantle, causing the liquid molten rock to move as it does so. This in turn causes the cooler magma beneath the crust to move along and then sink. This circular motion creates a constant 'current', which causes the tectonic plates floating on the mantle to move along with it.

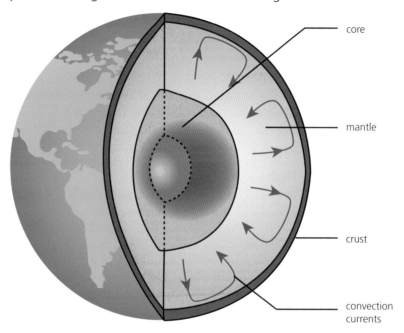

▲ **Figure 15.17** Convection currents in the mantle cause the tectonic plates to move.

DID YOU KNOW?

The study of earthquakes is called seismology. Scientists who study earthquakes are called seismologists, instruments that measure the strength of earthquakes are called seismometers and anything that is related to, or caused by, an earthquake is referred to as 'seismic'.

Science extra: Earthquake vibrations

When an earthquake occurs, it generates two kinds of seismic waves. Pressure waves (or **P waves**) can travel through liquid and solid, and when they pass from one to the other, they change their path. Second waves (or **S waves**) cannot travel through liquids and are reflected from a liquid surface. The structure of the inside of the Earth was worked out by locating where an earthquake occurred and then recording where the P and S waves returned to the surface by detecting their vibrations.

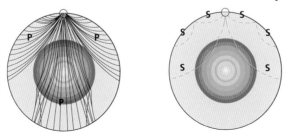

▲ **Figure 15.18** Patterns of P and S waves

A model Earth is used in the following enquiry. It is made from a tin and lid, which represent the Earth's crust, and an object inside the tin which represents materials vibrating as if a wave were passing through them.

Can you make deductions from vibrations?

You will need:

a small tin with a lid, and a range of small objects made from different materials collected by your teacher.

Hypothesis

As the knowledge of the structure and materials inside the Earth has been deduced by observing the vibrations of seismic waves, it should be possible to deduce the structure and type of materials inside a container by making the container vibrate.

Prediction

When a container with an object inside it is shaken, the structure and material of the object can be deduced.

Process

1. Ask your partner to put one of the items from your teacher in the tin and secure the lid. Do not look while your partner does this.
2. Take the tin and shake it as gently, forcefully, slowly or quickly as many times as you like and then make a deduction about the item inside. Write down what you think the item is.
3. Take out the item and compare it with your deduction.
4. Repeat steps 1–3 with all the items.

Examining the results

How many items were correctly deduced? What percentage of items were deduced correctly?

Conclusion

Compare your results to the hypothesis and prediction and draw a conclusion.

Is your conclusion limited in some way? Explain your answer.

What improvements could be made? Explain the changes that you suggest.

You can bring together different areas of science you have studied by making a moving model of tectonic plates. In the following enquiry, you will draw on your knowledge of densities, thermal convection and your ability to assemble equipment and use it safely.

Can you model tectonic plates on the move?

You will need:

a large beaker, a plain biscuit, a large tin of syrup, a Bunsen burner or other heat source, a tripod, gauze, a heat-proof mat.

Planning and investigating

Make a plan to show how two parts of a biscuit can be moved apart by convection currents. In your plan

a   describe how you will model the tectonic plates

b   describe how you will model the mantle beneath them

c   describe how you will model the heat from the Earth's core

d   describe how you will generate a convection current

e   show how you will assemble the equipment to make your moving model and how you will make your tectonic plates move

f   identify what you need to do to make sure you will be working safely as you are working with a heat source.

Show your plan to your teacher. If approved, make your model and demonstrate it under your teacher's supervision. Your heat source for the model should be a low blue flame and it should be directly under the centre of the beaker.

Conclusion

Explain how your knowledge of density and convection helped you to construct your model.

What are the analogies in your model?

What are the strengths and limitations of your model?

Summary

✔ Evidence for tectonic plates includes the following:
  – There are similarities between rocks found on the west coast of Africa and the east coast of South America.
  – The study of fossils indicates that continents may once have been joined together.
  – The location of volcanoes and earthquakes is directly related to the boundaries of tectonic plates.
  – Magnetic stripes created by minerals in newly created crust.
✔ Tectonic plates are moved by convection currents in the Earth's mantle.
✔ Knowledge about the movement of continents and the Earth's landmasses has been developed by many different people over a long period of time.

End of chapter questions

1 Which types of rock form under ground?
2 How are sedimentary rocks formed?
3 What was a mesosaurus?
4 How do scientists explain why the glossopteris plant grew in South America, South Africa, Australia and Antarctica?
5 How does plate movement cause an earthquake?
6 Describe the process of thermal convection in the mantle. You may draw a diagram if you wish.

 Now you have completed Chapter 15, you may like to try the Chapter 15 online knowledge test if you are using the Boost eBook.

Cycles on Earth

In this chapter you will learn:
- about the carbon cycle
- where Earth's carbon originally came from (Science extra)
- how photosynthesis, respiration, feeding, decomposition and combustion affect the carbon cycle
- how plants use glucose (Science extra)
- about the effect of carbon dioxide in our atmosphere
- the difference between weather and climate
- what climate change is and why most people think climates are changing
- the effects of climate change on our seas, atmosphere and weather
- about the global impacts of climate change (Science in context).

Do you remember?

- What is the rock cycle?
- What can eventually happen to the rocks that make up a mountain top?
- What is the water cycle?
- What happens to change seawater into a cloud?
- What do you understand by the word 'respiration'? Where does this process occur?
- What is the difference between a herbivore, a **carnivore** and an **omnivore**?
- What is the role of decomposers?

▲ **Figure 16.1** Ban Gioc waterfall, Cao Bang, Vietnam.

What is the carbon cycle?

From your studies of rocks and water, you know that they do not remain the same permanently. Water in a puddle can change into water vapour in the air on a hot day, while rocks on a mountain can take millions of years to become sand on the shore by the sea. One of the elements from which we are made also takes part in a cycle, and changes in its cycles can take from a fraction of a second to hundreds of millions of years. This element is carbon.

▲ **Figure 16.2** Carbon is cycling through this habitat at different speeds.

Carbon is the sixth element in the periodic table and has the symbol C. It can form chemical bonds with other elements, called covalent bonds (see Chapter 7). One atom of carbon can combine with up to four atoms of other elements using this type of bond.

The different processes in the carbon cycle can be seen in Figure 16.3. We will explore these processes in more detail in this chapter.

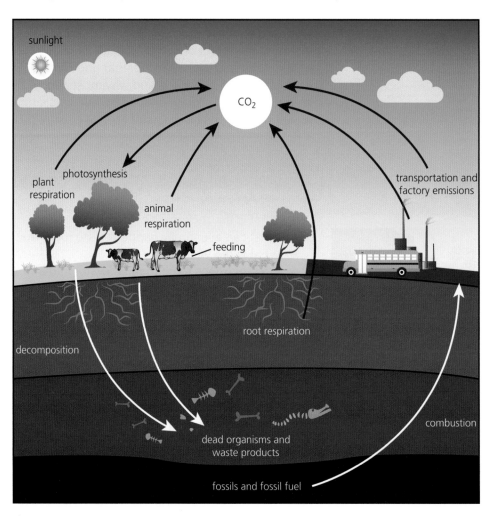

▲ **Figure 16.3** The carbon cycle.

Science extra: Where did the carbon on Earth come from?

When the **Big Bang** occurred 13.8 million years ago and produced the universe, two elements were formed: hydrogen and helium. Their atoms were pulled together by gravity to make stars, and in the stars, hydrogen was converted to helium which released energy as light and heat. When the stars used up their hydrogen, they began to make other elements, including carbon. In time, the stars stopped producing elements and released matter into the universe, slowly in the form of gases, or violently as in a supernova. Stars still behave in this way today.

When our Sun formed, it produced a disc of matter around it, which became the planets and other objects in the solar system. Some of the carbon in this disc formed the material in the Earth and this is still cycling through the solids, liquids and gases of our planet today.

Processes in the carbon cycle

Photosynthesis

Carbon is present in the air in the form of carbon dioxide. Most land plants have holes in their leaves, which allow gases produced in the leaf to pass through them into the air. These holes also allow gases, such as carbon dioxide in the air, to enter the leaf.

1 What is an endothermic reaction?

When sunlight shines on a leaf, the stomata open and carbon dioxide enters them. In the chloroplasts of the leaf cells, some of the energy in sunlight is used to produce an endothermic chemical reaction called **photosynthesis**. In this reaction, water from the soil and carbon dioxide from the air produce glucose and oxygen. The stomata of the leaf release oxygen and the glucose is converted to starch, which is stored in the leaf.

> ### Can you find the holes in a leaf?
>
> **You will need:**
>
> a plant leaf, a white tile or other flat surface, clear nail polish, clear sticky tape, a microscope.
>
> **Process**
> 1 Place the leaf on a flat surface with its lower surface upwards. Make sure it is flat and not curled up.
> 2 Brush the lower surface with clear nail polish and let it set. This will make a transparent cast of the surface of the leaf.
> 3 Place a piece of clear sticky tape over the cast and attach the cast to it.
> 4 Lift the tape and cast clear of the leaf and put it down on a microscope slide with the cast touching the glass. Press the tape to the slide to make the cast secure and cut away any pieces of tape which overhang the slide.

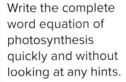

CHALLENGE YOURSELF

Write the complete word equation of photosynthesis quickly and without looking at any hints.

DID YOU KNOW?

The process in which plants take carbon from the atmosphere and use it to build molecules for living things is called **carbon fixation**.

5　Place the slide on the stage of the microscope and, starting with the low-power lens, look for the holes in the leaf cast. The hole is called a **stoma** and its plural is **stomata**, so you will be looking for stomata on the leaf. Each one is surrounded by two slightly curved cells which look a little like bananas or sausages. They are called **guard cells**.

6　If you think you have found some stomata, draw a few and show them to your teacher to check that you have found them.

Science extra: How plants use glucose

We have seen that glucose is converted into starch to be stored, but it is also used by plants in respiration, which is discussed in the following section. Glucose is used by plants in other ways as well, as Figure 16.4 shows.

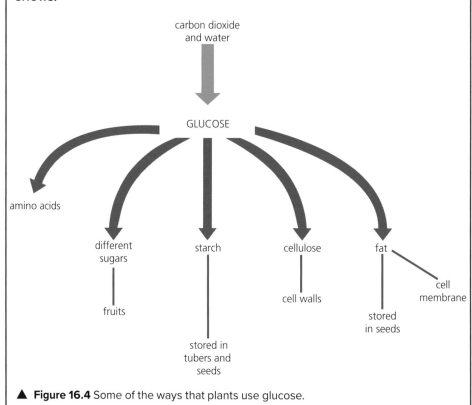

▲ **Figure 16.4** Some of the ways that plants use glucose.

Respiration

Respiration links to photosynthesis in the carbon cycle. In photosynthesis, energy is trapped in chemical compounds for use in living organisms. In respiration, this energy is released slowly and in a controlled way through a series of chemical reactions taking place in the mitochondria of cells. The process of respiration can be represented by the word equation:

 glucose + oxygen → carbon dioxide + water

The energy is used for all life processes and involves assembling molecules to make bodily structures, such as bones and muscles. It can also be converted to kinetic energy to allow organisms to move.

We tend to think that photosynthesis is the only major process taking place in plants, but respiration is taking place too. It can be demonstrated in plants that are not photosynthesising – when they are in their seed forms! The following investigation shows that indicators are not just used in chemistry but in biology investigations too.

Do germinating pea seeds produce carbon dioxide?

You will need:

some soaked peas, a piece of muslin and thread to tie into a bag, a gas jar, a cover sealed with Vaseline, hydrogen carbonate indicator solution in the gas jar, as shown in Figure 16.5.

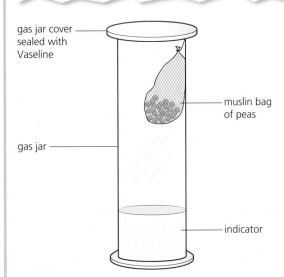

gas jar cover sealed with Vaseline

muslin bag of peas

gas jar

indicator

▲ **Figure 16.5** Investigating carbon dioxide production by germinating pea seeds.

Hypothesis

Complete this hypothesis:

Hydrogen carbonate indicator changes from red to yellow in the presence of carbon dioxide. If the peas are respiring ...

Prediction

Make a prediction about what you think will happen.

Process

1 Place the peas in the centre of the sheet of muslin and pull up the sides around them, twist, and tie the bundle closed with a thread.
2 Pour the indicator into the gas jar to the level shown in the diagram.
3 Hang the bag of peas into the jar, but pull the thread over the edge of the jar and place the gas jar cover over it.
4 Take a photograph of your equipment.
5 Decide on length of time to leave the equipment for, then return and photograph it again.
6 If you detect a slight colour change but are not sure, leave the equipment for a longer period of time, then return and photograph it again.

Analysis and evaluation

Arrange the photographs in time order and compare them.

Conclusion

Does your conclusion match your hypothesis and prediction? Have you collected enough evidence?

LET'S TALK

The equipment shown in Figure 16.5 could be used to show that maggots release carbon dioxide. Should animals be used in experiments to show signs of life, such as respiration?

When scientists read about experiments, it sometimes gives them ideas for investigative questions. Asking a question about the respiration rate of beans is an example.

Do all beans respire at the same rate?

You will need:

several sets of equipment as shown in Figure 16.5, samples of a variety of different kinds of beans (they do not have to be all the same size) which have been soaked for a day.

Hypothesis

Soaked beans begin to germinate. In this process, they respire to release energy.

Look at your samples of beans and construct a testable hypothesis based on the information in the sentences above and your observations on the different kinds of beans.

Prediction

Base your prediction on your hypothesis.

Planning, investigating and recording data

Construct a plan to investigate your hypothesis and prediction. In your plan, mention the variables involved, how you will control risks, the frequency of observations, the need for accuracy and precision, and the method of recording data. If your teacher approves your plan, try it.

Examining the results

Look through your data and identify any trends, patterns or anomalous results.

Conclusion

Compare your evaluation with your hypothesis and prediction and draw a conclusion.

Were any of the results unexpected? If so, how do they affect your understanding of respiration in the different beans?

How does the evidence you have collected support or refute your prediction?

What are the limitations of your conclusion? How could the investigation be improved in order to
a   provide more reliable data
b   build new experiments on what you have discovered?

Feeding

Plants use the carbon they take in during photosynthesis to make carbohydrates, fats and proteins, which they use to keep themselves alive. These substances also form the food for animals, which means that carbon then passes along food chains. The animals in a food chain can be described as **primary consumers** if they eat plants, **secondary consumers** if they eat primary consumers, and **tertiary consumers** if they eat secondary consumers.

plants → primary consumers → secondary consumers → tertiary consumers

2   Name two primary consumers.
3   Are herbivores primary consumers? Explain your answer.
4   Name two secondary consumers.
5   Are all secondary consumers carnivores? Explain your answer with examples.

6 How many food chains are linked to trees in Figure 16.6? Describe each one.

CHALLENGE YOURSELF

Make a drawing of a food chain that includes a rabbit, a fox and a plant. Use your drawing to plan a model using electrical components to show the passage of carbon along a food chain, using two cells, three lamps and as many wires and switches as you need. The model should show the passage of carbon along the food chain by lighting the lamps in sequence. Make a circuit diagram of your model and show how the lamps relate to the organisms in the food chain. If your teacher approves, make and demonstrate your model.

Food chains link together to form food webs, so there are many paths for carbon to cycle through in the organisms in a habitat, as Figure 16.6 shows.

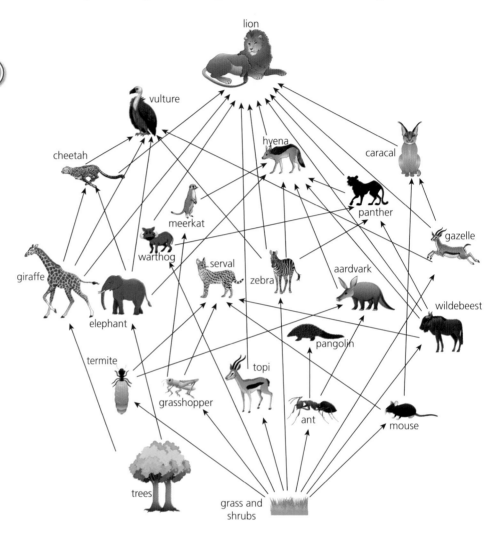

▲ **Figure 16.6** A food web of the savannah ecosystem in Africa.

Decomposition

When plants and animals die, their bodies become food for decomposers. Carbon in plant and animal bodies passes into the bodies of micro-organisms such as bacteria and fungi. It also passes into the bodies of invertebrates, such as earthworms and insect larvae like the maggot larvae of flies. As the decomposers feed, they release energy from their food in respiration and produce carbon dioxide which passes into the air.

▲ **Figure 16.7** Fungi decomposing a fallen tree.

▲ **Figure 16.8** Coal was formed millions of years ago in swamps like this one.

For decomposition to take place, the decomposers need to have oxygen around them that they can use in respiration. If this is lacking, the decomposers cannot feed and break down the dead bodies. This situation occurred in swamps millions of years ago, when tree-like plants fell into them and did not decompose. Instead, they formed the fossil fuel: coal.

In ancient seas, the dead bodies of microscopic marine organisms fell to the sea floor, where there was not enough oxygen for decomposers to survive and, in time, their bodies changed into another fossil fuel: oil.

Combustion

Humans have used **combustion** (burning of fuel) for almost half a million years to keep warm and to cook food. These processes are essential to life. When a fuel such as wood is set on fire, the carbon it contains combines with oxygen in the air to produce carbon dioxide and water. The word equation for this reaction is:

 fuel + oxygen → carbon dioxide + water

> **CHALLENGE YOURSELF**
> **Groupwork**
>
> How could you show that a burning candle produces carbon dioxide?
>
> Work in a group and discuss your ideas. Think back to prior learning and try to remember an experiment you might have done or seen. Set out what you would do and, if your teacher approves, try it.

Carbon dioxide the 'greenhouse gas'

Carbon dioxide is classified as a 'greenhouse gas' because carbon dioxide (and several other gases) in the atmosphere act like the glass in a greenhouse. Greenhouse gases allow heat energy from the Sun to travel through them and reach the surface of Earth, but then they prevent much of the heat energy radiating from the Earth's surface from passing back out into space, so the heat is retained. The heat energy remains in the atmosphere and warms it up, as Figure 16.9 shows.

CHALLENGE YOURSELF

Carbon dioxide is not the only greenhouse gas. Do some research to find out the names of some other greenhouse gases and how they are found in the atmosphere. What human activities cause other greenhouse gases to be released?

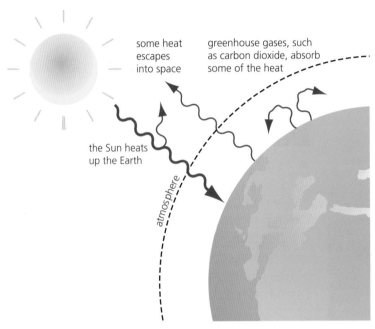

▲ **Figure 16.9** Some of the heat that the Earth receives from the Sun is trapped in the atmosphere through the natural greenhouse effect.

The natural warmth of the Earth has allowed millions of different life forms to develop, and it keeps the planet habitable. Without the natural greenhouse effect, humans, animals and plants would not be able to live on Earth.

How can you demonstrate the greenhouse effect?

Glass or plastic jars are made of the same materials as greenhouses. You can use a glass or plastic jar and two thermometers to demonstrate the greenhouse effect in a simple way. You choose a suitable location, somewhere that is warm, and record the temperature on both thermometers. Then you place one thermometer inside the jar with the lid on to seal it. At regular intervals for a fixed amount of time you take temperature recordings from both thermometers.

Plan

In your plan, state the equipment you will need and how you will use it. Explain how you will make a fair test, including the measurements you will take and the frequency at which you will take them to provide reliable data. Be sure to state where you will place your jars and equipment for the duration of your experiment.

Examining the results

Look for trends, patterns and anomalous results in your data. If there are anomalous results, can you suggest reasons for them?

Conclusion

Decide whether your investigation answers the question and draw a conclusion.

What are the strengths and limitations of your model?

How could you improve it to make the data more reliable?

7 How is the term 'weather' different from the term 'climate'?

Climate change

Climate is the average weather experienced at a particular place or area over a long period of time.

Different parts of the world have different climates. These are called **climate zones**. They range from the coldest climate zones at the poles of the planet to the hottest climate zones found closest to the equator. All the different climates depend upon certain features of the atmosphere, which have been measured for many years at weather stations on the planet and by satellites around it. These features include temperature, precipitation, humidity, wind and atmospheric pressure. These aspects of the climate affect the plants and animals that can live in the various climate zones and have been constant in those zones for about ten thousand years.

▲ **Figure 16.10** A weather satellite monitoring a tropical storm.

Why might climates be changing?

Just over two hundred years ago, the **Industrial Revolution** occurred in several countries of the world. Factories were built to manufacture a very wide range of products and all of those factories needed energy to provide power for the machines. The fuel that provides energy was originally coal. As more coal was burnt, it added more carbon dioxide to the atmosphere. As time went on, more and more factories were set up and more and more coal was burnt. Later, oil was also used for power and, as this was burnt even more, more carbon dioxide entered the atmosphere. Today, the burning of coal and oil continues in power stations around the world.

Scientists have measured the amount of carbon dioxide in bubbles trapped in ice at the poles which date back 800 000 years, and have linked this data with other ways of measuring carbon in more recent times. From this data, the graph in Figure 16.11 has been produced.

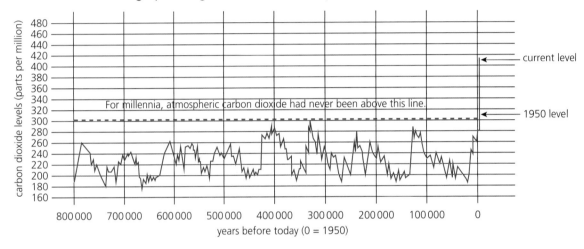

▲ **Figure 16.11** Carbon dioxide levels over time.

8 Does the graph in Figure 16.11 show a pattern or a trend? Explain your answer.

9 Does the graph in Figure 16.12 show a pattern or a trend? Explain your answer.

LET'S TALK

How would you use the greenhouse model to explain to someone about the danger of too much carbon dioxide in the planet's atmosphere?

Scientists have also measured the global temperature from 1880 up to the present day. This data has been constructed by taking the very different temperatures in different parts of the world and calculating a mean (average) for the whole planet.

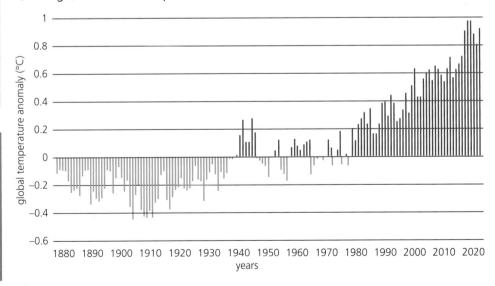

▲ **Figure 16.12** Global temperature since 1880.

The effects of climate change

Climate change describes the differences in measurements of temperature, precipitation, humidity, wind and atmospheric pressure over a long period of time. These differences are recorded around the world. Many people believe that these changes are making the Earth warmer. The term **global warming** describes this process.

There are several impacts of climate change which we will look at in closer detail.

Sea level change

As global temperatures rise, ice at the poles, in glaciers and on mountains melts and the water flows into rivers, seas and oceans. This causes sea levels to rise and low-lying islands just above sea level can be totally flooded. Some islands have already been lost and, if the sea levels rise further, even more islands could disappear.

10 If sea levels continue to rise, will it affect the area where you live? Explain your answer.

▲ **Figure 16.13** Islands such as the Maldives are at great risk of flooding.

The coastlines of large land-masses are also affected by rising sea levels. As water creeps further up the shore at high tide, it floods the land, forcing people who live in coastal areas to move further inland.

Extra water in the atmosphere

Rising temperatures cause the rate of evaporation to increase. This means that the air holds more water and this disrupts the flow of air around the planet, changing weather patterns almost everywhere around the world.

In some places, changes in weather patterns produce periods of very heavy rain, which cause rivers to fill quickly and cause flash floods that can destroy the surrounding land. In other places, changing weather patterns produce longer periods of drought, which can destroy habitats and farmers' crops and livestock. There have always been floods and droughts, but they may be happening more often in recent times, and people are worried that they might happen even more often in the future.

11 What problems are caused by long periods of drought?

LET'S TALK

What problems do you think people living in coastal areas face as a result of flooding? Why do you think people are being forced to move further inland? In a group, discuss the possible problems caused by flooding, then share your ideas with the class.

Extreme weather events

A **hurricane** (also called a tropical storm or **cyclone**) is an example of an extreme weather event. Hurricanes form over the water of warm seas. As global temperatures rise, more water evaporates to form the hurricane, so they are carrying more water than before. When hurricanes hit land, they lose energy and release this water, causing floods. Remaining water in the hurricane falls as torrential rain.

LET'S TALK

Are there rivers in your country which can flood the surrounding land if there is a lot of rain? What are the problems caused by these floods? Think about people who live in the area, their crops and livestock, and also the wild plants and animals. Discuss your ideas in a group.

▲ **Figure 16.14** Flooding caused by hurricane Irma in 2017.

Extreme weather events such as hurricanes cause great damage and are a threat to life. Other examples include **tornadoes** and heavy snowstorms (blizzards).

Scientists are recording the number of extreme weather events that happen around the world each year. They want to know if these are occurring more often. If they are, this might be because of global warming.

▲ **Figure 16.15** Tornado.

▲ **Figure 16.16** Snowstorm.

CHALLENGE YOURSELF

Use the internet to find out about extreme weather events in your country. Do you think they are becoming more or less frequent? What data can you collect to support your ideas?

CHALLENGE YOURSELF

Use the internet to find out about how scientists use climate models. Make a presentation of your discoveries by producing a poster, fact file, mind map or an illustrated written account.

CHALLENGE YOURSELF

Look at the map in Figure 16.17. Use an atlas, globe or the internet to identify some of the small islands that are at risk of flooding. Some of these islands could disappear completely within your lifetime.

Use the internet to do some further research about what is being done to reduce the impact of flooding in these areas, and how citizens of those places feel about the future of their island homes. Write a report or give a short presentation on your findings.

Remember: When you use the internet for research, think carefully about your sources of information!

Science in context

Climate change

We have seen how climate change is having a major impact around the world due to changing sea levels, rising temperatures and the increased frequency of extreme weather events. These things may appear unrelated and you may think that, in isolation, if they only happen in certain places then the worldwide impact can't be so great. However, Figure 16.17 shows the true magnitude of the effects of climate change on our planet. As we can see here, climate change truly is a global issue.

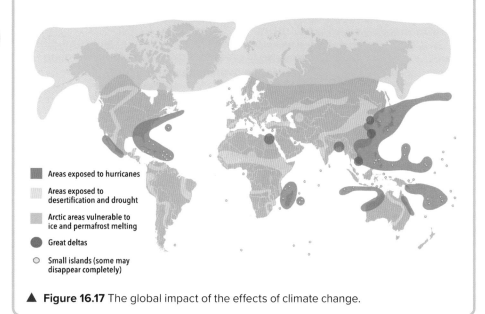

- ■ Areas exposed to hurricanes
- ▨ Areas exposed to desertification and drought
- ■ Arctic areas vulnerable to ice and permafrost melting
- ● Great deltas
- ○ Small islands (some may disappear completely)

▲ **Figure 16.17** The global impact of the effects of climate change.

DID YOU KNOW?

Deltas are landforms found at the end of rivers, where they meet the sea. Deltas are areas where sediment (material carried by the river) is deposited and builds up over time, because the sea is not strong enough in that location to carry it away. So deltas are never found above sea level. Climate change presents a problem for deltas as sea levels continue to rise, which puts deltas at risk. Deltas are home to a huge variety of plants and animals, especially fish. They are also home to many people who rely on the delta for fishing or farming.

LET'S TALK

Work in a group. Look at the great deltas marked on the map in Figure 16.17. Use an atlas, globe or the internet to identify one or more of them, and the country in which they are found. What effects of climate change present the most risk in the area you have chosen? What might happen to the delta in future? Why is it important for people in this area to protect the delta, and what could they do? Discuss your ideas and listen to the ideas of others.

Remember: When you use the internet for research, think carefully about your sources of information!

CHALLENGE YOURSELF

Many governments are looking at ways of reducing climate change. Which ones could you and your friends help with? How could you change the way you live in order to slow down global warming? What changes do you think you will have to make in the future?

Use the internet or talk to people you know to find out what things people can do about climate change, and write a list of ideas. Share your ideas in a group and discuss them.

Remember: It is always important to remain positive and look for solutions which will help people cope with the changes, or which will prevent further global warming.

Summary

- ✔ The way that carbon passes between the atmosphere and living things can be represented by the carbon cycle.
- ✔ In photosynthesis, plants remove carbon dioxide from the atmosphere to create glucose.
- ✔ All living things respire and the process can be represented by the equation:

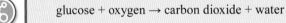

$$\text{glucose} + \text{oxygen} \rightarrow \text{carbon dioxide} + \text{water}$$

- ✔ Carbon dioxide produced through the processes of respiration and combustion is added to the atmosphere.
- ✔ Carbon passes through the bodies of organisms in a food chain when they feed.
- ✔ When living organisms die, decomposers release carbon dioxide into the atmosphere.
- ✔ It is predicted that climate change will alter sea levels, thus increasing flooding in some places and causing drought in others.
- ✔ Extreme weather events such as hurricanes, tornadoes and extreme snowfall may be more powerful and more frequent due to climate change.

End of chapter questions

1 Describe the passage of carbon up a food chain.
2 What happens to carbon in the process of combustion?
3 Why are some gases, like carbon dioxide, referred to as 'greenhouse' gases?
4 Name an extreme weather event and describe how it is affected by climate change.
5 How is climate change affecting the place where you live and what are people doing about it?

6 How could the activities of people on the planet change to reduce global warming and climate change? If you wish, you may answer in the form of a science-fiction story, as sometimes science fiction can become science fact.

 Now you have completed Chapter 16, you may like to try the Chapter 16 online knowledge test if you are using the Boost eBook.

In this chapter you will learn:

- where asteroids are found in the solar system
- how asteroid collisions with Earth can lead to climate change and mass extinctions
- how new species can emerge after a mass extinction event (Science extra)
- how asteroid collisions with Earth may have led to the formation of the Moon
- where stars are born
- about astronomy across the world (Science in context)
- about the unsolved mysteries of the universe (Science extra).

▲ **Figure 17.1** The Earth in space.

Do you remember?

- Name the planets of the solar system, starting with the one nearest the Sun.
- What is a galaxy?
- What is a planetary system?

Asteroids

Where are asteroids found?

Earth and other planets and objects in space were formed from the disc of dust and gases around the Sun, but not all the material there formed planets and moons. Some rocky objects remained much smaller and formed **asteroids**. Most asteroids are found in a ring around the Sun, between the orbits of Mars and Jupiter. This ring is known as the **asteroid belt**. There are two other groups of asteroids, called **trojans**. They are found on either side of Jupiter and move in its orbit around the Sun. The positions of the asteroid belt and the trojans in the solar system are shown in Figure 17.2 on the next page.

> **DID YOU KNOW?**
> There are millions of asteroids in the asteroid belt, but they cover such a large area of space that there can be as much as 515 000 km between any two asteroids!

CHALLENGE YOURSELF

Some large asteroids have been given names. Many names relate to characters from ancient stories and legends. Use the internet to find out the names and some other facts about asteroids. Try to make a note of at least ten facts that you discover. Then share and compare your findings in a small group.

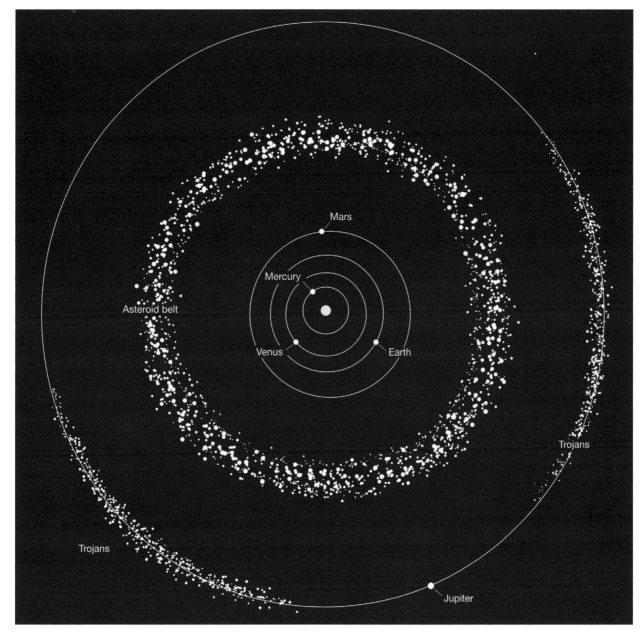

▲ **Figure 17.2** Positions of asteroid belts and trojans in the solar system.

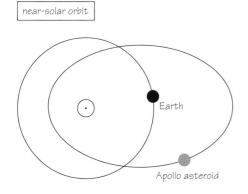

near-solar orbit

Earth

Apollo asteroid

Asteroid collision

Asteroids are held in position by the forces of gravity between them and the forces of gravity between them and Jupiter. As all these objects move in their orbits, there can be changes in the forces of gravity, which can cause an asteroid to change the direction of its movement and set up another orbit around the Sun. Some of these orbits may cross the orbits of planets and their moons and, in time, may lead to a collision.

◀ **Figure 17.3** The path of an asteroid across the Earth's orbit.

When an asteroid collides with a planet or a moon, it produces an impact **crater**. On our Moon, where there is no atmosphere to cause weathering and erosion, the craters remain as they first formed after impact. The Moon keeps one surface facing towards Earth as it moves through space; this is called the near side and it is the side that we see. The other side, the side we never see, is called the far side.

a   b

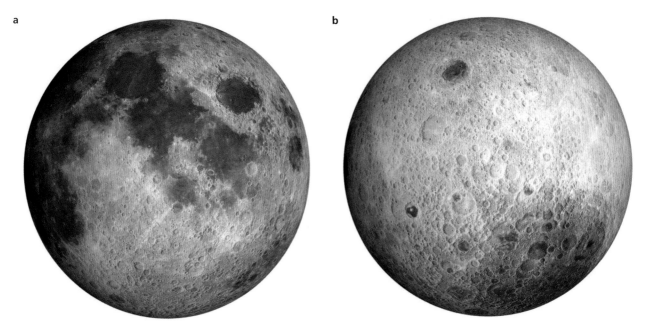

▲ **Figure 17.4** The craters on **a** the near and **b** the far side of the Moon.

1 Look at the craters on both sides of the Moon. How do they compare? Construct a hypothesis to explain what you see.

How does asteroid speed affect crater depth and width?

You will need:

a bowl of flour, sand or moist clay to represent the surface of the Moon, a model of an asteroid (attach a thread to it, so that you can pull it straight up out of the crater after impact), a wooden spill on which millimetre marks have been made (to measure crater depth), a ruler (to measure crater width), a metre rule (to measure 'asteroid' height above the 'surface of the Moon').

Hypothesis

Construct a testable hypothesis to answer the question.

Prediction

Make a prediction based on your hypothesis.

Planning and investigating

Construct a plan to investigate your prediction, explaining

a   your choice of asteroid and Moon surface

b   your experimental technique

c   the number of measurements taken and how they are recorded

d   what risks need to be considered.

If your teacher approves your plan, try it.

Examining the results

Present your data in the form of a graph and predict results between the data points collected. In the data, look for a trend or a pattern, then identify any anomalous results and explain them.

Conclusion

Compare your evaluation with your hypothesis and prediction and draw a conclusion.

What are the limitations of your conclusion? How might the conclusion be further investigated?

What improvements could you make to your investigation? Explain why you would make them.

The effects of asteroid collision on Earth

Climate change

The climate in any place on the planet depends on the amount of heat and sunlight it receives from the Sun. When an asteroid hits the land on Earth, it sends a huge amount of dust and ash into the atmosphere. If it lands in the sea, it also sends up water vapour and droplets. This means that wherever the asteroid strikes, it sends material into the atmosphere which blocks out the Sun's light and heat. This results in a darkening of the skies and a cooling of the planet's surface. These conditions are sometimes described as an 'impact winter' and can last for a number of years.

Mass extinctions

A mass extinction is an event in which a very large number of plant and animal species (more than 75 per cent of the species on the planet) become extinct. They die out because their populations shrink to a size where the death rate is greater than the reproductive rate. The reason for their death is a change in conditions on the Earth.

There are many sources of change in conditions on the Earth. The activities of a large number of volcanoes erupting at the same time, for example, can produce changes in the atmosphere when they emit ash and smoke, which block out sunlight. The planet also varies its tilt as it orbits the Sun, and this can reduce the amount of heat being received in the northern hemisphere, which can start an ice age. Such changes can lead to extinctions of plant and animal species.

There have been five mass extinctions in Earth's history and scientists have been unsure of their causes, but the last one, which happened 66 million years ago, is thought to have been brought about by an impact winter. This was called the Cretaceous–Paleogene extinction and was responsible for the extinction of the tyrannosaurus rex and many other land-dwelling dinosaurs, as well as other animal, reptile and marine species, such as ammonites.

▲ **Figure 17.5** An artist's impression of the asteroid in the atmosphere at the end of the age of dinosaurs.

> **DID YOU KNOW?**
> Many scientists believe that today we are living through the sixth mass extinction event, and that the main cause of the loss of animal and plant life is human activity on the planet.

It is believed that when this happened, there was a great deal of volcanic activity on the Earth. An asteroid of between 11 and 81 kilometres in diameter struck the Earth and made a huge crater, now called the Chicxulub crater, which is 20 kilometres deep and can be found in Mexico. It is believed that the impact would have produced instant **firestorms** and complete devastation over thousands of square kilometres around it before the impact winter set in.

2 How do you think an impact winter affects plants?

3 How do you think an impact winter affects carnivores?

4 Use the internet to find the location of the Chicxulub crater on a map, and the location where you live. If the impact happened today, would it affect you? Explain your answer.

▲ **Figure 17.6** The asteroid colliding with Earth.

Science extra: New species

After a mass extinction, many species that shaped habitats and upon which other species relied are wiped out. This presents an opportunity for other species that survived the extinction to evolve into new species, as conditions become more favourable for them. Mammal species which survived mass extinctions have evolved into a wide range of species, including humans. Dinosaur species which had feathers and survived the last mass extinction evolved into birds, so when you hear a bird singing, you are listening to a living dinosaur!

The formation of the Moon

There is a theory for how the Moon formed. It is a collision theory called the giant-impact hypothesis, but is also known as the **Big Splash**. Scientists have constructed this collision theory by studying a variety of topics, such as the rocks on the Moon and the Earth, the way the Earth and Moon move, and observations and theories about the material that forms around stars. Evidence to support the hypothesis comes from a variety of sources.

Evidence from rocks

The composition of rocks on the Moon and the Earth is identical, suggesting that they formed from one rocky body in space. Moon rock samples show evidence of being molten at some time, which would happen in a collision.

Evidence from movement

The spin of the Earth and the way the Moon orbits the Earth suggest there is a close relationship between them.

Evidence from observations and theories

Stars have been found with discs of material around them, called debris discs, which could have formed by objects colliding. The theory of the formation of the solar system supports the idea of objects in the early solar system colliding together.

The collision theory of the Moon's formation has been developed from this evidence and suggests that, about 4.5 billion years ago, the Earth that was forming crashed into a planet-like object about the size of Mars, which scientists called Theia. This collision resulted in a ring of rocky debris forming around the Earth from the two objects. In time, gravitational forces pulled the rocks in the ring together to form the Moon. In the collision, Theia was destroyed.

▲ **Figure 17.7** The Big Splash.

Can you model the Big Splash?

You will need:

modelling clay, video recording equipment (optional).

Planning and investigating

Show how you could break up a lump of modelling clay and use it to model how the young Earth and Theia came together and produced a debris ring that eventually became the Moon.

You may wish to make a storyboard before you start making the model.

If you can, you could make a recording of how your model can be used to explain the Big Splash.

Where stars are born

There are huge clouds of dust and gas spread out across the universe. These clouds are called **nebulae**. A nebula may form from the gas and dust produced when a star explodes in a supernova. It may also be developed as gravity pulls together the gas and dust in intergalactic space. In this type of nebula, the forces of gravity between the atoms and molecules of helium and hydrogen gas bring them together into a huge ball. The force of gravity continues to act between the gas particles and squashes them closer and closer together, and the pressure inside the ball increases. Eventually, the pressure pushing on the hydrogen atoms becomes so great that they fuse together to make helium atoms and release energy as heat and light. Gradually, the ball begins to glow and eventually, as the pressure continues to increase and more hydrogen is converted into helium, the ball starts to shine and becomes a star. The star is then said to be 'born'. As this 'birthing' process takes place inside nebulae, they are called **stellar nurseries** and many stars can be forming inside them at the same time.

▲ **Figure 17.8** A stellar nursery.

Can you make an electric stellar nursery?

You will need:

one or more cells, lamps, switches, a variable resistor, a number of wires depending on your model, translucent material, video recording equipment.

Planning and investigating

Use the equipment to show how one or more stars in a stellar nursery start to glow and shine.

Make one or more circuit diagrams to make a star shine. Make a drawing of how you will use the translucent material to make a nebula.

Show your work to your teacher and, if approved, make your stellar nursery.

Science in context

Astronomy across the world

Even the earliest humans would have looked up at the night sky and wondered at it. As time went on, human civilisations began to explain the night sky in a variety of ways and related it to events taking place on the Earth. Today we are still fascinated by the night sky, and scientists study it to find out more about it and our place in the universe. This in turn helps us try to understand to existence, and for many, this means trying to conserve the conditions on Earth which sustain life by maintaining ecosystems and trying to control climate change.

Astronomy is the study of objects found in space, such as stars, planets, moons and asteroids. It is also the study of phenomena that occur in space, like the formation of stars, the movement of galaxies and the formation of planets. Scientists who study astronomy are called astronomers and many of them work in observatories around the world, collecting data from huge telescopes which use light or radio waves to look into space. There are major observatories all around the world – some of the biggest can be found in the USA, the Canary Islands and Chile. Most observatories are built away from cities, where the sky is not as polluted, so a clearer view of space can be seen.

▲ **Figure 17.9 a** Roque de Los Muchachos Observatory, La Palma, Canary Islands, Spain and **b** Mauna Kea Observatories, Big Island, Hawaii, USA.

CHALLENGE YOURSELF

Find the location of the nearest observatory to where you live. It might be a major observatory like those in Figure 17.9 or it may be a smaller, local observatory. Do some research about the observatory and the astronomers who work there.

Scientists and astronomers also collect data from telescopes on satellites and instruments on space probes, which visit planets and moons in the solar system. All the data collected by observatories from satellites is recorded and examined, and then conclusions can be drawn.

CHALLENGE YOURSELF

In prior learning, you may have found out about exoplanets and how they were discovered. If you did, look at your sources again and see how many more planets have been discovered since then. What do you find?

CHALLENGE YOURSELF

Use the internet to see if you can find out the latest ideas about dark matter and dark energy. Do scientists agree or do they hold different views and make different explanations?

Remember: When you use the internet for research, think carefully about your sources of information!

Ideas, conclusions and suggestions from scientists are published as research papers and read by other scientists around the world. This process is called **peer review**. In the process of peer review, comments about the data and constructive criticisms of the methods of observation are made, so that a general conclusion can be drawn and agreed on. As we have seen in science, a conclusion is only provisional and may have to be revised when other observations and experiments are made.

Observatories around the world also scan the skies for asteroids approaching the Earth (known as **Near-Earth Objects**) and when one is discovered, the data about it is shared by all observatories worldwide. This allows further observations in a variety of locations to be made and the risk of a collision or threat can be calculated more accurately. In this way, astronomy helps to keep the planet and its inhabitants safe.

Science extra: Understanding the universe

The universe is still a mysterious place for scientists. The laws of physics that explain how objects move cannot explain how galaxies rotate. If the laws applied, the galaxies would pull themselves apart. As they do not do this, scientists believe that there is some matter that cannot be seen helping the galaxies to rotate. Scientists have called this **dark matter**, but they do not know what it is.

The movement of galaxies presents another mystery that cannot be explained: they are moving away from each other much faster than expected as the universe continues to expand. The source of the energy which is causing this movement is unknown, and scientists call this **dark energy**. Discoveries about dark matter and dark energy are constantly being made in science.

Summary

- ✔ An asteroid can collide with planets and moons.
- ✔ The effects of an asteroid collision with Earth include climate change and mass extinctions.
- ✔ It is thought that our Moon was formed when Earth and another small planet collided. The small planet was destroyed as a result of the collision.
- ✔ Nebulae are huge clouds of dust and gas that are spread out across the universe. The force of gravity between particles in these clouds forms stars in stellar nurseries.
- ✔ Astronomers around the world collect data from observatories and satellites to increase their understanding of the universe. New theories are evaluated by other scientists in a process known as peer review.

End of chapter questions

1 What did the disc of gas and dust around the Sun form?
2 What holds the asteroids in place?
3 What makes an asteroid change its orbit?
4 What can happen on Earth if a large asteroid crashes into it?
5 How did the Moon form?

6 Why is astronomy important to us today?

Now you have completed Chapter 17, you may like to try the Chapter 17 online knowledge test if you are using the Boost eBook.

Science today and tomorrow

What kind of scientist might you be?

At the beginning of this science course, in *Student's Book 7*, you were asked what kind of scientist you might like to be, and you were shown a range of scientific equipment. At the start of *Student's Book 8* you were challenged to explain how to use this equipment and think creatively when trying to solve scientific problems. At the start of this book, you were shown how the signs of being a scientist develop as a person grows up, and you were asked to assess your own personal signs of being a scientist.

▲ **Figure 1** Students in a science lab.

LET'S TALK

What kind of a scientist did you think of being at the start of this course? If you were to go on and be a scientist, what kind would it be now?

Are you scientifically literate?

Throughout your learning, you have carried out science activities such as planning an enquiry, taking measurements, drawing conclusions and taking part in peer reviews.

Scientists are also very critical of their own work and the work of others. When they are examining models, they consider their strengths and weaknesses; if they are drawing conclusions, they look for how these conclusions may be limited by the experiment, and how making improvements may help to find out more.

These activities give a 'hands on' experience of doing science, which help you understand how scientists work to make their discoveries. From these discoveries we build up knowledge and understanding of the four scientific subjects considered in the course: biology, chemistry, physics and Earth and space.

▲ **Figure 2** Scientists in discussion.

Being scientifically literate, then, means knowing a range of facts about the topics in each of the subjects – such as photosynthesis for biology and electricity for physics – and also knowing about the different scientific activities that scientists use to make their discoveries.

Even if you do not go on to be a scientist, you need to take these features of scientific literacy with you. The reason for this is that during your life you will be faced with making decisions about science-based issues – such as tackling infectious diseases, using energy and materials efficiently, conserving habitats, plants and animals and looking at ways to reduce the effects of climate change. This information will be presented to you in newspapers, TV programmes and internet platforms.

When you study this information, remember how scientists work – they collect lots of data, they make models that have strengths and weaknesses and their work must be peer reviewed. After you have assessed how the information provides you with these details (or not), you can then decide how to respond to it. You are behaving as a scientifically literate person.

Can you explain things with science?

Another, perhaps more fun, way of assessing your scientific literacy is to look at an everyday scene and explain scientifically what you see. Here is an example – the campfire scene shown in Figure 3.

▲ **Figure 3** A campfire.

LET'S TALK

Even if you do not plan to become a scientist, there may be some scientific topics that you have found interesting and may like to read or watch films about in the future. What might these topics be?

LET'S TALK

What other relationships can you describe scientifically in the picture? For example, what is the relationship between respiration in the people around the fire and the plants surrounding them?

You could begin with the logs on the fire and explain that during photosynthesis many years ago, the Sun's energy was stored in the plant, which in turn stored some of this energy in the form of wood. The burning of the wood is releasing this energy as heat, which is moving out through radiation to warm the faces of the people around the fire, and passing by conduction through the cooking pot, and by convection in the liquid in the food that is cooking.

The theory of everything

If you moved away from the campfire and continued describing the world in scientific terms, you may feel that you are developing a 'theory of everything'. You would not be the first to do this. The Ancient Greeks tried to explain everything from their observations in terms of four elements that they devised – earth, fire, water and air. This theory was popular for many centuries. Democritus' ideas about atoms as a theory of everything did not become popular until the time of the work of John Dalton and others.

Theories of everything try to link up everything we know and understand about science in a simple way. For example, Galileo's work on how things fall and Kepler's work on how planets move in their orbits were brought together by Newton's studies on gravity and his law of universal gravitation. All theories of everything depend on the work of scientists through history.

Science moving forwards

You have looked at science and where you are now, but what about the future?

Science is always moving forwards. You have seen through 'hands on' work that one investigation can lead to another, and that later investigations reveal even more knowledge and understanding than those that went before. Those discoveries help us live our lives.

The application of scientific knowledge in this way is called technology, and the applications themselves – from sewage works to smart phones – are examples of technology.

As science moves forwards, so technology moves forwards too. The main technological needs to help us live our lives now are: medicines and equipment to maintain our health, devices to use energy efficiently and reduce climate change, processes for recycling materials, improved agricultural practices to grow enough food for everyone, equipment for monitoring habitats to help conserve biodiversity and to reclaim habitats that have been lost due to human activities.

In addition to scientific knowledge and understanding, we need creativity and hard work to apply technology to help us in our lives. A great example

CHALLENGE YOURSELF

In the *Checkpoint Science Student's Books* for Stages 7, 8 and 9, over a hundred scientists have been featured. Try to name ten of them and briefly describe what they did.

DID YOU KNOW?

The James Webb Space Telescope (JWST) was launched into space on Christmas day, 2021. The JWST is an example of a large and complicated piece of technology which will further our understanding of the universe. Search the internet to learn more about it.

of this is the development of The Great Green Wall in Africa, which, when completed, will be 8000 kilometres long and 15 kilometres wide, running across the continent. It is being constructed by planting drought-resistant trees, using low-tech items such as ploughs and hoes, but their use is preventing the development of a desert, and will instead provide land on which people can live.

LET'S TALK

Creativity is a key feature of science. People can become totally absorbed in their studies, but it is also important to take time to relax. Perhaps you may like to celebrate the completion of your Cambridge Checkpoint Science course by using your creativity to organise a science-themed party.

Here are some ideas to consider before you set up the party:

- Should it be indoors or outdoors?
- Should it have a theme (everyone coming dressed as atoms and molecules, or as scientists from the past, perhaps, or should the theme be open to any science topic you have studied, so that you can be surprised at the range of ideas and costumes?)?
- Can you think of a way of including science-themed food and drink?
- What about choosing science-themed music and inventing a dance?

Whatever you do, remember to take some time to relax from your studies. You will find that it helps to keep you well, and, while relaxing, you may have a creative thought which leads to finding out more about our world and its place in the universe.

Glossary

A

Acid A substance with a pH less than 7 that reacts with metals to produce hydrogen.

Adaptation The features of a living thing which are suited to its habitat so that it can survive there.

AIDS The letters used to describe acquired immune deficiency syndrome which is caused by an infection of the human immunodeficiency virus, called HIV. This infection damages the immune system of the body and can be fatal.

Amino acids Molecules made mainly from atoms of nitrogen, carbon, oxygen and hydrogen which bond together to form larger structures, called proteins, which form structures in the cells to keep them alive.

Ammeter An instrument for measuring the size of a current flowing through a circuit, in amperes.

Amplitude The maximum height of a crest or depth of a trough from the rest position which is half-way between them.

Analogy A comparison of one thing which is unfamiliar or difficult to understand with something else which is more familiar and easier to understand. The aim of the comparison is to make understanding easier.

Anion An ion with a negative electrical charge.

Anther The organ in a flower that produces pollen grains.

Asteroid Rocky objects smaller than planets and moons that orbit the Sun, mainly between the orbits of Mars and Jupiter.

Atom A particle from which all substances are made. For example, diamond is made up of carbon atoms.

Atomic number The number of protons in the nucleus of an atom.

B

Big Bang An explosive event that scientists believe produced the universe.

Big Splash A hypothesis to explain the formation of the Moon by a collision between the Earth and a planet-like object called Theia.

C

Carbohydrate A nutrient made from carbon, hydrogen and oxygen. Most are made by plants. Examples are glucose, starch and cellulose.

Carnivore An animal that feeds only on other animals.

Catalyst A substance that speeds up a chemical reaction without being changed itself or used up in the reaction.

Catalytic converter A device attached to the exhaust pipe of petrol and diesel engines which contains a catalyst involved in chemical reactions which convert the harmful chemicals in the exhaust gases into less harmful ones.

Cation An ion with a positive electrical charge.

Characteristic An observable feature that is always found in a particular type of organism.

Chromosome A thread-like structure that appears when the cell nucleus divides. It contains DNA.

Combustion A chemical reaction in which a substance reacts quickly with oxygen and heat is given out in the process. If a flame is produced, this is called burning.

Concentration The word concentration is used to describe the amount of a substance that is present in a certain volume of a mixture. It is used when referring to the amount of a solute (something dissolved in a liquid) dissolved in a certain volume of a solution.

Conduction The transfer of heat energy from one part of a solid material to another by vibrating particles in the material. Conduction also refers to the ability of material to allow electrons to pass through it in the form of a current of electricity.

Conductor A material that allows electricity or heat to pass easily through it, by conduction.

Conservation In biology the word means the care and protection of populations of organisms in their habitats so that they will be able to contribute to the biodiversity of the planet long into the future. The word conservation is also used to describe the protection of resources such as energy sources and minerals for future generations of humans. In chemistry and physics the word refers to the fact that matter or energy are not created nor destroyed but simply change forms.

Consumer An animal that eats either plants or other animals.

Continuous variation The way a feature in a species varies by small amounts over a wide range.

Convection The transfer of heat energy through a liquid or a gas by the movement of particles in the substance.

Covalent bond The bond formed between atoms, by the atoms sharing the electrons in their outer shells.

Crater A large, deep, round-shaped hollow on the surface of the Earth, the Moon, other planets or asteroids, produced by a collision with an object large enough to make an impact. Or, in the case of craters on Earth, the centre of a volcano collapsing.

Craton An area of the Earth's crust covered in sedimentary rock which has not been disturbed by actions such as mountains forming or volcanoes for a very long time (in billions of years).

Creativity Being able to develop imaginative ideas in a variety of ways in the form of pictures, objects or processes such as new experiments.

Curiosity Wanting to know about something.

Current A flow of electrically charged particles. The unit in which it is measured is the ampere (amp). The unit symbol is A.

Cyclone A very large storm with very high winds that form in the South Pacific and Indian Oceans.

D

Decibel A unit used in the measuring of the loudness of a sound. The Unit symbol is dB.

Decomposers Organisms such as soil bacteria, fungi, earthworms and other invertebrates which feed on the bodies of dead plants and animals and reduce them to their minerals which then enter the soil.

Density The mass (amount of matter) of a substance that is found in a certain volume. A high-density substance contains a large mass in a small volume. A low-density substance contains a much smaller mass in the same volume.

Discontinuous variation The way a feature in a species varies distinctly from one individual to the next.

Displacement The process in which one thing is replaced with another. For example, in a displacement reaction, a metal in a salt is replaced by another metal.

Displacement-time graph A type of graph which shows the different positions particles can occupy when a sound wave is produced.

DNA The letters stand for deoxyribonucleic acid. Chromosomes contain genes that contain DNA which determines how a particular characteristic, such as hair colour or eye colour, will develop in an organism.

E

Ecosystem An ecological system in which the different species in a community interact with each other and with the non-living environment. Ecosystems are found in all habitats, such as lakes and woods.

Electron A tiny particle inside an atom which moves around the nucleus. It has a negative electric charge.

Endothermic reaction A chemical reaction in which heat energy is taken in.

Environment The surroundings or conditions in which an organism lives.

Enzyme A chemical made by a cell that is used to speed up chemical reactions in life processes such as digestion and respiration.

Erosion The transport of small rock fragments by air and water from one location to another.

Ether A colourless liquid composed of carbon, hydrogen and oxygen atoms used as a solvent in industry and an anaesthetic in medicine.

Evaporation A process in which a liquid turns into a gas without boiling.

Exothermic reaction A chemical reaction in which heat energy is released.

F

Fertilisation The fusion of the nuclei from the male and female gametes that results in the formation of a zygote (fertilised egg).

Fertiliser A substance that is added to the soil to maximise plant growth.

Firestorm A storm produced by a large fire which draws in air from all around it, creating a wind that increases the burning process and brings in yet more air to continue burning up the trees where it began.

Food chain A series of organisms linked together by the passage of food between them. When a food chain is drawn as a diagram, arrows show the direction of the energy flow through the system.

Food web The way a number of food chains in a habitat link together to show how food and energy pass through the habitat.

Force A push or pull that may start or stop an object moving, or change the direction or speed of movement. It can also make an object change shape and size.

Fossil A body, body part or trace of a plant, animal or other organism that has become preserved, usually in stone.

Frequency The number of waves passing a point in a certain amount of time. It is measured in Hertz.

Fuel Material such as coal, oil, gas or wood that is burned to release its chemical energy, producing heat and light.

G

Gamete A sex cell, which contains half the number of chromosomes in a normal body cell of the species. During reproduction, a female gamete and a male gamete fuse at fertilisation, forming a zygote (fertilised egg) with the full number of chromosomes.

Gene A section of DNA that contains the information about how a particular characteristic, such as hair colour or eye colour, can develop in the organism.

Giant covalent structure A structure, called a lattice, that is produced by large numbers of atoms bonded together by covalent bonds.

Global warming Increase in average temperature of the Earth, which is likely to lead to significant climate changes.

Greenhouse gas A gas in the atmosphere that absorbs some of the heat energy from the sun with the result that the atmosphere becomes warmer.

Group (of periodic table) A column of elements in the periodic table. They have common properties, which show a trend down the group.

H

Habitat The home of a plant or animal – the place where it lives.

Heat dissipation A spreading out of heat from a place of high heat to many places of lower heat.

Herbicide A chemical that kills plants.

Herbivore An animal that feeds only on plant material. In food chains, herbivores are primary consumers.

Hertz A unit used in the measurement of the frequency of a sound wave or electromagnetic wave. The unit symbol is Hz.

Humidity A description of the amount of water vapour in the air. A large amount is described as high humidity, such as in a tropical rainforest and a small amount as low humidity, such as in a desert.

Hurricane A very large storm which very high winds that form in the North Atlantic and North East Pacific oceans.

Hypothermia A fall in the body temperature, accompanied by shivering at first, but later by a loss of consciousness which can lead to death if the body is not carefully warmed up again.

I

Imagination Being able to think up new ideas about what is being studied.

Industrial Revolution A change in the making of goods in small quantities by hand in small workshops, to making them in very large quantities by machines in factories.

Infrared radiation A form of energy in the form of electromagnetic waves which cannot be seen but can be felt as heat.

Inheritance In biology, the passing on of genes from one generation to the next.

Insecticide A chemical that kills insects.

Insulator A material that does not let heat or electricity pass through it. In heat insulators, the particles do not pass the energy from one to the next. In electrical insulators, the electrons are not allowed to flow and produce a current of electricity. Good insulators are also known as bad conductors.

Ion An atom which becomes electrically charged due to a change in the number of electrons it has around its nucleus. A group of atoms can also form an ion.

J

Joulemeter A piece of equipment that measures energy in units called joules.

K

Key A series of statements which can be used with observations of an organism to identify it.

L

Laser A word made up from its description – Light Amplification by Stimulated Emission of Radiation. The word is applied to the equipment that produces a narrow beam of light with all the light waves arranged so they have enough power to break up materials in a controlled way such as cutting metal.

Levity An imagined property of matter in which a substance was so light in weight that it would float upwards if it was not held down.

Loudness A measure of how quiet or loud a sound is. It is measured in decibels and is related to the amplitude of the sound wave.

M

Magma The molten rock beneath the Earth's crust.

Maize A crop plant that was originally grown in Central America, but is now grown in many parts of the world. A maize plant is a type of grass and produces many seeds packed tightly together in a long cylindrical-type structure known as an ear or a cob.

Mass The amount of matter in a substance and also in a certain volume of a substance.

Meteorological The scientific study of the atmosphere to help in forecasting the weather.

Microchip A tiny electrical circuit, called an integrated circuit, set on a very small, thin piece of silicon which controls one or more operations taking place in a larger electrical circuit.

Muslin A cotton cloth in which the threads are lightly woven together so that there may be some spaces between them through which air or water may pass.

N

Natural selection The process by which evolution is thought to take place. Individuals in a species best suited to a particular environment will thrive there and produce more offspring, while less well-suited individuals will produce fewer offspring. In time the less well suited will die out, leaving the best-suited individuals to form a new species.

Near-Earth Objects Objects in the solar system that are in orbit around the Sun, such as asteroids and comets but have paths which cross the orbit of the Earth, bringing the Earth and the object close together with a possibility of a collision.

Nebula (*plural:* nebulae) A huge cloud of gas and dust in space.

Neonatal The time period from the birth of a baby until it is 28 days old.

Neutron A particle in the nucleus of an atom that has no electrical charge.

Nucleus (*plural:* nuclei) The control centre of a living cell, containing the genetic material, DNA. In Chemistry, it refers to the positively charged centre of an atom.

O

Omnivore Animal that feeds on both plants and animals.

Orbit The curved path of one object around another, such as an electron around the nucleus of an atom, or of a moon around a planet.

P

Parallel circuit A circuit in which some components are set side by side and the current flows in parallel paths through them.

Parasitic The description of an organism that lives on or in the body of another organism and obtains all its food from it.

Patience Being able to wait for something to happen, or to continue working on a project which has difficulties in the hope it will be successful.

Period (of periodic table) The elements in each row of the periodic table.

Perseverance Being able to keep working on a project despite difficulties, instead of giving up.

Pesticide A chemical substance used to kill pests (for example, insects) that would otherwise damage crops.

Photochemical smog A type of air pollution in which the polluting chemicals from traffic react in sunlight to produce ozone and other chemicals in a haze which can cause damage to the respiratory system.

Photosynthesis The process by which plants make food (sugar) and oxygen from water and carbon dioxide, using energy from light that has been trapped in chlorophyll.

Pitch A measure of the frequency of a sound wave.

Pollen grain Microscopic grains produced by the anther, which contain the male gamete for sexual reproduction in flowering plants.

Pollination The transfer of pollen grains from an anther to a stigma, for sexual reproduction in flowering plants.

Population The number of individuals of a species living in a habitat.

Precipitate A solid substance which forms when a change takes place in a solution.

Precipitation In studies on the weather, this word refers to the falling of water through the atmosphere in various forms such as rain, snow or hail. In chemistry it means the formation of a solid precipitate in a liquid.

Predator An animal that feeds on other animals.

Pregnant A word used to describe the condition when a woman is having a baby. In biology it is also known as the period of gestation which is common to all female mammals having young developing inside them.

Pressure The term used to describe a force acting over an area of known size.

Prey An animal that is fed on by another animal.

Primary consumer The first animal in a food chain, which feeds on plants. Herbivores are primary consumers.

Producer The plant in a food chain, which makes food by photosynthesis, using energy from sunlight.

Products The substances (solids, liquids and gases) produced as a result of a chemical reaction.

Proton A particle in the nucleus of an atom that has a positive electrical charge.

Q

Quaternary consumer The fourth consumer along a food chain, feeding on a tertiary consumer. Quaternary consumers are always predators, but rarely prey animals, since food chains do not usually contain more than five organisms (including the producer).

R

Radiation The transfer of energy from one place to another by electromagnetic waves such as those of light or infrared radiation (heat).

Rate A measure of the speed at which something changes or takes place.

Reactants The substances (solids, liquids and gases) that take part in a chemical reaction.

Reactivity series The arrangement of metals in order of their reactivity with oxygen, water and acids, starting with the most reactive metal.

Resistance The property of a material that opposes the flow of an electric current through the material. The unit in which resistance is measured is the ohm. The unit symbol is Ω.

Resistor A component of a circuit that offers a certain amount of resistance to a current passing through a circuit.

Respiration The process occurring in all living organisms in which energy is released from food inside cells. Glucose reacts with oxygen to release energy for life processes, and carbon dioxide and water are produced. (Not be confused with breathing, which is the process of moving air in and out of the body.)

S

Salt A compound that is formed when an acid reacts with a substance such as a metal, or a metal carbonate. A salt is also formed in a neutralisation process between an acid and an alkali.

Satellites A spacecraft which moves in orbit around a planet, moon or the Sun to collect data and send it to Earth. It can also mean a moon which orbits a planet.

Scientific Revolution An event which took place in the sixteenth and seventeenth centuries, when a new way of investigating scientifically, using the scientific method was devised.

Secondary consumer The second consumer along a food chain, feeding on a primary consumer. Secondary consumers are always predators, and often also prey animals, being fed on by tertiary consumers.

Series circuit A circuit in which electrical components are arranged in a line and the current flows through them one after the other.

Simple structures Structures that are made from a small number of atoms that are held together by covalent bonds.

Sound energy The movement of energy in the form of sound waves through a solid, liquid or a gas.

Space probe A space craft without a crew, which carries instruments for measuring a wide range of things, such as temperatures and cameras for taking pictures, and sending the data back from all parts of the solar system to Earth.

Species The smallest group in a biological classification system. All the organisms it contains are of the same type. Individuals of a species have a large number of similarities and are able to reproduce with each other to form offspring but are not able to breed with different species.

Starch A carbohydrate made from glucose molecules joined together in a chain. It is the energy storage molecule for plants.

Stigma The part of a plant where pollen grains are trapped.

Subatomic particles The tiny particles from which atoms are made – protons, neutrons and electrons.

Superconductivity A property a material has if it offers no resistance to an electrical current.

Superconductor A material that does not offer any resistance to the flow of electricity through it. This property might be achieved when the material is made very cold.

T

Tectonic plate A huge slab of rock with an irregular shape that forms a large portion of the Earth's crust.

Tendril A long thin structure which grows out from a plant and coils round a support to help the stem grow upwards.

Temperature This is a measure of the hotness or coldness of a substance.

Tertiary consumer The third consumer along a food chain, feeding on a secondary consumer. Tertiary consumers are always predators, and occasionally also prey animals, being fed on by quaternary consumers.

Thermal energy transfer The process of moving thermal energy (heat) from one place to another.

Tornado A very fast rotating column of air produced beneath a funnel-shaped cloud. The wind moves so fast it rips up homes and trees and lifts them into the air, along with people and animals. Tornados occur on most continents and the largest numbers happen in the United States, Bangladesh and Argentina.

Transpiration The process by which plants lose water from their leaves.

V

Variation Differences between individuals in a species – for example, in humans, there is variation in hair colour and eye colour and height.

Vibrations Rapid up and down or side-to-side movements of an object or the particles in a substance.

Voltage The difference in electrostatic potential energy (the potential difference) between two points, such as the terminals of a cell. It is a measure of the ability of a cell to drive a current of electricity around a circuit. The unit in which it is measured is the volt. The unit symbol is V.

Voltmeter A device that measures the difference in electric potential between two parts of a circuit, in volts.

W

Wavelength The distance from the top of a wave's crest to the top of the next wave crest, or from the bottom of one wave trough to the bottom of the next wave trough.

Weathering The breaking down of the surface of rocks into smaller fragments.

Weight The downward force on an object due to gravity.

X

Xylem The name of a tube in a plant which conducts water from the root to the leaves.

Z

Zygote A fertilised egg cell.

Index

Acknowledgements

With warm thanks to Judith Amery, Kate Crossland-Page and Jennifer Peek for their contributions to this title.

Photo credits

p.vi © BQ-Studio.ru/stock.adobe.com; **p.viii** *t* © NYPL/Science Source/Science Photo Library; **p.ix** © Dadang Tri/REUTERS PHOTOGRAPHER/Alamy Stock Photo; **p.x** © Cozine/stock.adobe.com; **p.xi** © Sheila Terry/Science Photo Library; **p.xii** Courtesy of Women in Science website; **p.xiv** © Neville Elder/Corbis/Getty Images; **p.xv** *t* © DC Studio/stock.adobe.com; **p.xv** *b* © Erwin Sparreboom/Shutterstock.com; **p.xvi** © Rawpixel.com/Shutterstock.com; **p.2** © Abrilla/stock.adobe.com; **p.3** *b* © Tim UR/stock.adobe.com; **p.7** *l* © Olya/stock.adobe.com; **p.7** *c* © RJ22/Shutterstock.com; **p.7** *r* © Aleksa/stock.adobe.com; **p.8** © DigitalGlobe; **p.11** © Khajornkiat Limsagul/Shutterstock.com; **p.14** *l* © Wayne Lawler/Science Photo Library; **p.14** *c* © Peter Chadwick/Science Photo Library; **p.14** *r* © Gregory A. Pozhvanov/Shutterstock.com; **p.16** *t* © Science Photo Library; **p.16** *b* © Science Photo Library; **p.17** © Brian Gadsby/Science Photo Library; **p.22** © John Keates/Alamy Stock Photo; **p.25** © Xinhua/Shutterstock; **p.26** © Markobe/stock.adobe.com; **p.29** © James King-Holmes/Science Photo Library; **p.32** © Kateryna_Kon/stock.adobe.com; **p.34** © Tim Vernon/Science Photo Library; **p.35** *t* © Science Source/Science Photo Library; **p.35** *b* © A. BARRINGTON BROWN, © GONVILLE & CAIUS COLLEGE / SCIENCE PHOTO LIBRARY; **p.36** *t* © Jim Sugar/The Image Bank/Getty Images; **p.36** *b* © Ron Frehm/AP/Shutterstock; **p.38** *t* © Sheila Terry/Science Photo Library; **p.38** *b* © Vermontalm/stock.adobe.com; **p.39** *t* © Sheila Terry/Science Photo Library; **p.39** *b* © Kevin Schafer/Alamy Stock Photo; **p.43** © Tchara/stock.adobe.com; **p.44** *t l* © Eric Isselee/Shutterstock.com; **p.44** *t r* © WildMedia/stock.adobe.com; **p.44** *b* © Prostock-studio/stock.adobe.com; **p.45** © KostyaK/stock.adobe.com; **p.48** *t* © Nenov Brothers/stock.adobe.com; **p.48** *b* © Paulaphoto/stock.adobe.com; **p.50** *t* © Ingehogenbijl/Shutterstock.com; **p.50** *b, l* © Kokliang1981/stock.adobe.com; **p.50** *b, r* © PAOLO/stock.adobe.com; **p.51** *t* © Sanjay Thakur/Ephotocorp/Alamy Stock Photo; **p.51** *b* © Pedro/stock.adobe.com; **p.52** © Bruno Petriglia Science Photo Library; **p.57** © Jeremy Cozannet/Alamy Stock Photo; **p.58** *t* © Alexey Masliy/Shutterstock.com; **p.58** *b* © Granger/Shutterstock; **p.61** © Artegorov3@gmail/stock.adobe.com; **p.62** © Ksena32/stock.adobe.com; **p.63** © FLHC 2021BB/Alamy Stock Photo; **p.65** *t* © INTERFOTO/Personalities/Alamy Stock Photo; **p.65** *b* © Science Museum Library/Science & Society Picture Library -- All rights reserved.; **p.68** *l* © Dmytro/stock.adobe.com; **p.68** *c* © Romaset/stock.adobe.com; **p.68** *r* © Science Photo Library/Alamy Stock Photo; **p.74** © Alessandro/stock.adobe.com; **p.75** © Sam Ogden/Science Photo Library; **p.76** © Tibor Bognar/Alamy Stock Photo; **p.78** © KiwiCo - www.kiwico.com **p.82** © Feng Yu/stock.adobe.com; **p.90** © Redkphotohobby/stock.adobe.com; **p.91** © Amir Cohen/REUTERS/ Alamy Stock Photo; **p.93** © Andrew Lambert Photography/Science Photo Library; **p.95** © Andrew Lambert Photography/Science Photo Library; **p.97** © Alvey & Towers Picture Library/Alamy Stock Photo; **p.98** *t* © Andrew Lambert Photography/Science Photo Library; **p.98** *b* © Andrew Lambert Photography/Science Photo Library; **p.101** *t* © Sunny Forest/stock.adobe.com; **p.101** *b* © Andrew Lambert Photography/Science Photo Library; **p.102** © Andrew Lambert Photography/Science Photo Library; **p.103** © Martyn F. Chillmaid; **p.105** © Tapui/stock.adobe.com; **p.107** *t* © Kanjana/